GeoPlanet: Earth and Planetary Sciences

More information about this series at http://www.springer.com/series/8821

Jerzy Sasiadek
Editor

Aerospace Robotics III

Editor
Jerzy Sasiadek
Department of Mechanical and Aerospace Engineering
Carleton University
Ottawa, ON, Canada

The GeoPlanet: Earth and Planetary Sciences Book Series is in part a continuation of Monographic Volumes of Publications of the Institute of Geophysics, Polish Academy of Sciences, the journal published since 1962 (http://pub.igf.edu.pl/index.php).

ISSN 2190-5193 ISSN 2190-5207 (electronic)
GeoPlanet: Earth and Planetary Sciences
ISBN 978-3-319-94516-3 ISBN 978-3-319-94517-0 (eBook)
https://doi.org/10.1007/978-3-319-94517-0

Library of Congress Control Number: 2018945868

This Springer imprint is published by the registered company Springer Nature Switzerland AG
The registered company address is: Gewerbestrasse 11, 6330 Cham, Switzerland

Series Editors

Managing Editor

Anna Dziembowska
Institute of Geophysics, Polish Academy of Sciences

Advisory Board

Kon-Kee Liu
Institute of Hydrological
and Oceanic Sciences
National Central University Jhongli
Jhongli, Taiwan

Teresa Madeyska
Research Centre in Warsaw
Institute of Geological Sciences
Warszawa, Poland

Stanisław Massel
Institute of Oceanology
Polish Academy of Sciences
Sopot, Poland

Antonio Meloni
Instituto Nazionale di Geofisica
Rome, Italy

Evangelos Papathanassiou
Hellenic Centre for Marine Research
Anavissos, Greece

Kaja Pietsch
AGH University of Science and
Technology
Kraków, Poland

Dušan Plašienka
Prírodovedecká fakulta, UK
Univerzita Komenského
Bratislava, Slovakia

Barbara Popielawska
Space Research Centre
Polish Academy of Sciences
Warszawa, Poland

Tilman Spohn
Deutsches Zentrum für Luftund
Raumfahrt in der Helmholtz
Gemeinschaft
Institut für Planetenforschung
Berlin, Germany

Krzysztof Stasiewicz
Swedish Institute of Space Physics
Uppsala, Sweden

Ewa Szuszkiewicz
Department of Astronomy
and Astrophysics
University of Szczecin
Szczecin, Poland

Roman Teisseyre
Department of Theoretical Geophysics
Institute of Geophysics
Polish Academy of Sciences
Warszawa, Poland

Jacek Tronczynski
Laboratory of Biogeochemistry
of Organic Contaminants
IFREMER DCN_BE
Nantes, France

Steve Wallis
School of the Built Environment
Heriot-Watt University
Riccarton, Edinburgh
Scotland, UK

Wacław M. Zuberek
Department of Applied Geology
University of Silesia
Sosnowiec, Poland

Piotr Życki
Nicolaus Copernicus Astronomical
Centre
Polish Academy of Sciences
Warszawa, Poland

Preface

Dear Colleagues and Friends,

The *III Conference on Aerospace Robotics* (*CARO'2015*) took place in Warsaw in November 2015. We received large number of papers from institutions working on space robotics around the world. The accepted papers were presented over two days at the Space Research Institute Conference Centre.

It is with great pleasure that we would like to show once more to the international space robotics community selected papers presented at the *III Conference on Aerospace Robotics* (*CARO'2015*) held in Warsaw, Poland on 17–18 November 2015.

We have selected 11 papers presenting different aspects of aerospace robotics. The selected papers cover the broad range of space robotics including free-floating, flying and mobile robots but also some aspects of navigation.

We hope that our readers will find these papers not only interesting but also helpful and inspiring in their professional activities.

We would like to thank the organizers of the CARO'2015 from Space Research Centre PAN (CBK PAN), Military University of Technology and Warsaw University of Technology, in particular organizing committee, all plenary speakers and reviewers for their outstanding work.

Warsaw, Poland
September 2017

Iwona Stanisławska
Jerzy Sasiadek
Centrum Badan Kosmicznych PAN (CBK PAN)

Contents

Parallel Hamiltonian Formulation for Forward Dynamics of Free-Flying Manipulators

Paweł Malczyk, Krzysztof Chadaj and Janusz Frączek

1 Introduction

Free-flying space robotic systems in which manipulators are mounted on a spacecraft have been proposed in the field of space robotics to increase the mobility of such devices. In a number of space applications multi-arm robotic systems are exploited in order to broaden the manipulability properties of the system. Thus naturally, the dynamics, planning and control algorithms received significant attention from researchers (Umetani and Yoshida 1989; Vafa and Dubowsky 1990; Papadopoulos and Dubowsky 1991; Dubowsky and Papadopoulos 1993). One should also note that apart from internal joint degrees of freedom (DOF), free-flying manipulators have six additional DOF, usually assigned to the manipulator base-body (spacecraft). The additional mobility properties for such systems change a way of studying the dynamics and developing control algorithms for free-flyers. The complexity of the free-flying robotic systems increase and many practical maneuver realizations are dependent on the efficiency of computational algorithms (Yokokohji et al. 1993; Abiko and Hirzinger 2008). Various formulations for the dynamics of free-flying manipulators exist in the literature that help to plan and control multiple tasks and posture behaviors. Some of them are based on the operational space dynamics (Chang and Khatib 2000). Usually, efficient, linear time, recursive algorithms are used in order to predict the dynamics of such branching systems (Featherstone 1983; Jain and Rodriguez 1995). Nevertheless, the applications of these algorithms are somehow limited to sequential computations, which are performed along independent branches of the system.

P. Malczyk (✉) · K. Chadaj · J. Frączek
Division of Theory of Machines and Robots, Faculty of Power and Aeronautical Engineering, Institute of Aeronautics and Applied Mechanics, Warsaw University of Technology, Nowowiejska 24, 00-665 Warsaw, Poland
e-mail: pmalczyk@meil.pw.edu.pl

J. Sasiadek, *Aerospace Robotics III*, GeoPlanet: Earth and Planetary Sciences,
https://doi.org/10.1007/978-3-319-94517-0_1

On the other hand, with the advances in parallel computer architectures, researchers paid more attention to the development of parallel algorithms for multibody dynamics (see Malczyk and Frączek 2008 for the more detailed literature review). Featherstone (1999) developed truly optimal-time, logarithmic order divide and conquer algorithm (DCA) for the dynamics of tree-like topologies as well as closed-loop multibody systems (MBS). Recently, the divide and conquer schemes attract significant attention to the development of efficient parallel algorithms for large MBS, partially due to the fact that computationally powerful multicore processors or graphics processor units are cheaply available on the market. Various algorithms that are effectively based on the Featherstone's DCA are elaborated with a myriad of extensions to e.g.: humanoid robot simulations (Yamane and Nakamura 2009), molecular dynamics simulations (Malczyk and Frączek 2015), real-time applications, discontinues changes in the system topology, sensitivity analysis, constraint enforcement (Malczyk and Frączek 2012; Mukherjee and Malczyk 2013a) or nonholonomic constraints (Mukherjee and Malczyk 2013b). Advances in the application of the DCA approach to large multibody system simulations can be found in Laflin et al. (2014).

The most efficient rigid-body dynamics algorithms encountered in the literature are formulated with the aid of Newton-Euler's formalism (Jain and Rodriguez 1995). On the other hand, Hamilton's canonical equations constitute an interesting alternative as a basis for elaborating low order formulations (Lankarani and Nikravesh 1988; Naudet et al. 2003). As Hamilton's canonical equations are expressed in terms of velocities and momenta of the system and the possible constraints are usually imposed at the velocity level, many sources indicate profitable characteristics of the Hamiltonian approach and prospect for new insight into the dynamics. Surprisingly, the literature review reveals that it is difficult to find a fully parallel approach for multi-rigid body dynamics simulations based on the Hamiltonian approach. In the previous papers (Chadaj et al. 2015, 2017a) the authors proposed the Hamiltonian based divide and conquer formulation (HDCA) for the simulation of open-loop multi-rigid body dynamics. This paper discusses efficient way to model the dynamics of free-flying systems possessing tree topologies. A novel HDCA algorithm achieves logarithmic complexity in terms of the number of bodies in the system, and is linear, when the calculations are performed sequentially. The application of the HDCA algorithm for the dynamics simulation of multi-arm robotic system indicate the usefulness of the approach and the efficiency of the HDCA can be easily observed, especially for long kinematic chains as demonstrated by parallel performance results on graphics processor units.

2 Hamiltonian Based Divide and Conquer Formulation

In this section the Hamiltonian based divide and conquer algorithm for forward dynamics of free-flying manipulators possessing tree-like topologies will be presented. Although the state of the system is described by utilizing joint coordinates, the underlying dependencies are described with the use of absolute coordinates. The momentum conservation principle for the articulated and accumulated momenta allows one to determine the system's velocities and impulsive constraint load at joints in the assembly–disassembly manner achieving logarithmic computational cost for parallel programming. Afterwards, the equations of motion are formulated in terms of articulated momenta derivatives.

2.1 Joint Velocities

The spatial momentum vector $\mathbf{P}_O$ at arbitrary point O of a body (see Fig. 1) is defined as

$$\mathbf{P}_O = \mathbf{M}_O \mathbf{V}_O = \begin{bmatrix} m\mathbf{I} & -m\tilde{l}_{OC} \\ m\tilde{l}_{OC} & \mathbf{J}_O \end{bmatrix} \begin{bmatrix} \mathbf{v}_O \\ \boldsymbol{\omega} \end{bmatrix} \tag{2.1}$$

where $\mathbf{M}_O \in \mathcal{R}^{6\times6}$ is a mass matrix (m mass of the body, $\mathbf{J}_O$ moment of inertia) and $\mathbf{V}_O$ is a spatial velocity vector at point O consisting of translational $\mathbf{v}_O$ and angular $\boldsymbol{\omega}$ velocities (cp. Naudet et al. 2003). A skew-symmetric associated with the position vector l_{OC} is designated by a tilde symbol above the vector. As presented in the Fig. 1, let us define a rigid body A connected with the remaining bodies in the multibody system by kinematic joints at handles locations. Following the definition presented in Featherstone (1983), a handle is an abstract point at which the body communicates with the proceeding or subsequent body through the kinematic joint. Typically, the existence of the handle introduces further dependences on the momentum and force equilibrium of the investigated body.

Let us define the spatial articulated momentum vector $\bar{\mathbf{P}} \in \mathcal{R}^6$ related to a handle and consisting of linear and angular momenta as

$$\bar{\mathbf{P}} = \mathbf{D}\boldsymbol{\sigma} + \mathbf{H}\mathbf{p} \tag{2.2}$$

where the matrices $\mathbf{H} \in \mathcal{R}^{6\times n_f}$ and $\mathbf{D} \in \mathcal{R}^{6\times(6-n_f)}$ represent the joint's motion subspace and the subspace associated with the joint constrained directions, respectively (n_f—degrees of freedom for the joint). These subspaces are orthogonal, so the condition $\mathbf{D}^{\mathrm{T}}\mathbf{H} = \mathbf{0}$ is fulfilled. The quantities $\mathbf{p}$ are joint canonical momenta,

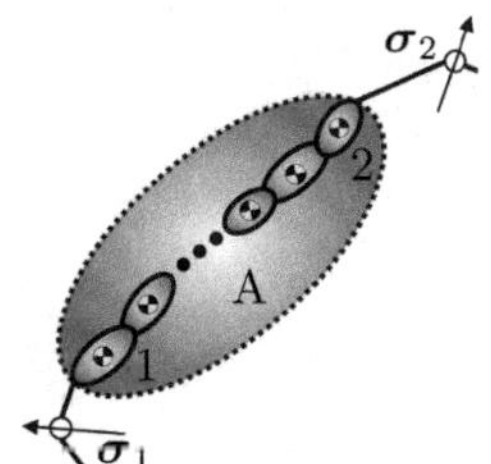

Fig. 1 A compound body A

and $\boldsymbol{\sigma}$ are constraint impulsive loads. The momentum conservation principle for considered body A with two handles at points O_1 and O_2 can be constructed as

$$\mathbf{M}_1^A \mathbf{V}_1^A = \overline{\mathbf{P}_1^A} + \mathbf{S}_{12}^A \overline{\mathbf{P}_2^A} \tag{2.3}$$

$$\mathbf{M}_2^A \mathbf{V}_2^A = \mathbf{S}_{21}^A \overline{\mathbf{P}_1^A} + \overline{\mathbf{P}_2^A} \tag{2.4}$$

Equation (2.3) refers to the point O_1, while Eq. (2.4) to the point O_2 of body A (cp. Fig. 1). Matrices $\mathbf{S}_{12}^A$ and $\mathbf{S}_{21}^A$ are shift matrices as introduced by e.g. Jain and Rodriguez (1995) from point O_1 to O_2 and from O_2 to O_1, respectively. The shift matrices facilitate transformations of spatial momentum, force and velocity vectors. To reveal the influence of constraints impulsive forces $\mathbf{T}_1^A$ and $\mathbf{T}_2^A$ on the velocities, the Eqs. (2.3)–(2.4) are reformulated into a concluding form of articulated momentum conservation principle for two-handle body

$$\mathbf{V}_1^A = \xi_{11}^A \mathbf{T}_1^A + \xi_{12}^A \mathbf{T}_2^A + \xi_{10}^A \tag{2.5}$$

$$\mathbf{V}_2^A = \xi_{21}^A \mathbf{T}_1^A + \xi_{22}^A \mathbf{T}_2^A + \xi_{20}^A \tag{2.6}$$

Now, consider momentum conservation principle of body B (Fig. 2)

$$\mathbf{V}_1^B = \xi_{11}^B \mathbf{T}_1^B + \xi_{12}^B \mathbf{T}_2^B + \xi_{10}^B \tag{2.7}$$

$$\mathbf{V}_2^B = \xi_{21}^B \mathbf{T}_1^B + \xi_{22}^B \mathbf{T}_2^B + \xi_{20}^B \tag{2.8}$$

The goal is to construct a divide and conquer assembly procedure to obtain the momentum conservation equations for compound body C consisting of bodies A and B in the form

$$\mathbf{V}_1^C = \xi_{11}^C \mathbf{T}_1^C + \xi_{12}^C \mathbf{T}_2^C + \xi_{10}^C \tag{2.9}$$

$$\mathbf{V}_2^C = \xi_{21}^C \mathbf{T}_1^C + \xi_{22}^C \mathbf{T}_2^C + \xi_{20}^C \tag{2.10}$$

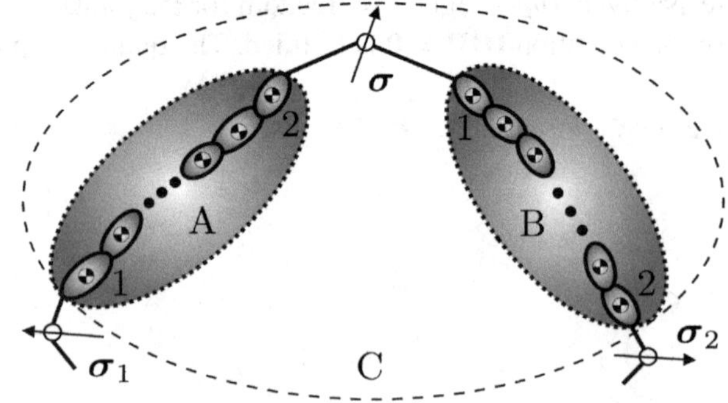

Fig. 2 Assembly of bodies A and B into C

where the handles O_1 of A and O_2 of B are coincident with handles of body C. In other words, the terms $\xi_{11}^C, \xi_{12}^C, \ldots, \xi_{20}^C$ have to be expressed in terms of known, corresponding terms for bodies A and B. The kinematic constraints for joint connecting A and B is

$$\mathbf{V}_1^B - \mathbf{V}_2^A = \mathbf{H}\dot{\mathbf{q}} \tag{2.11}$$

where $\dot{\mathbf{q}}$ is joint velocity. Substituting Eqs. (2.6) and (2.7) into Eq. (2.11) projected onto the joint constrained directions results in

$$\mathbf{D}^T\left(\xi_{11}^B\mathbf{T}_1^B + \xi_{12}^B\mathbf{T}_2^B + \xi_{10}^B - \xi_{21}^A\mathbf{T}_1^A - \xi_{22}^A\mathbf{T}_2^A - \xi_{20}^A\right) = \mathbf{0} \tag{2.12}$$

As the Newton's law has to be satisfied, the following condition has to be fulfilled

$$\mathbf{T}_1^B = -\mathbf{T}_2^A = \mathbf{D}\boldsymbol{\sigma} \tag{2.13}$$

The substitution of Eq. (2.13) into (2.12) allows us to resolve for the impulsive Lagrange multipliers $\boldsymbol{\sigma}$ as a function of external impulsive forces $\mathbf{T}_1^A$, $\mathbf{T}_2^B$

$$\boldsymbol{\sigma} = -\left[\mathbf{D}^T\left(\xi_{11}^B + \xi_{22}^A\right)\mathbf{D}\right]^{-1}\mathbf{D}^T\left(\xi_{12}^B\mathbf{T}_2^B - \xi_{21}^A\mathbf{T}_1^A + \xi_{10}^B - \xi_{20}^A\right) \tag{2.14}$$

If the constraints imposed on the system are not dependent, the inversion exists as inverted matrix is symmetric and positive definite. Let us define the quantities

$$\mathbf{W} = \mathbf{D}\mathbf{C}\mathbf{D}^T = -\mathbf{D}\left[\mathbf{D}^T\left(\xi_{11}^B + \xi_{22}^A\right)\mathbf{D}\right]^{-1}\mathbf{D}, \quad \boldsymbol{\beta} = \mathbf{W}\left(\xi_{10}^B - \xi_{20}^A\right) \tag{2.15}$$

to express the Eq. (2.14) as

$$\mathbf{T}_1^B = \mathbf{W}\xi_{12}^B\mathbf{T}_2^B - \mathbf{W}\xi_{21}^A\mathbf{T}_1^A + \boldsymbol{\beta} = -\mathbf{T}_2^A \tag{2.16}$$

The above Eq. (2.16) describes how constraint impulsive forces at external handles depend on the constraint impulsive forces at the joint connecting A and B. The substitution of Eq. (2.16) into Eqs. (2.5) and (2.8) allows us to obtain the coefficients of momentum conservation principle for articulated body C

$$\xi_{11}^C = \xi_{11}^A + \xi_{12}^A\mathbf{W}\xi_{21}^A, \quad \xi_{22}^C = \xi_{22}^B + \xi_{21}^B\mathbf{W}\xi_{12}^B \tag{2.17}$$

$$\xi_{12}^C = -\xi_{12}^A\mathbf{W}\xi_{12}^B = \left(\xi_{21}^C\right)^T, \quad \xi_{10}^C = \xi_{10}^A - \xi_{12}^A\boldsymbol{\beta}, \quad \xi_{20}^C = \xi_{20}^B + \xi_{21}^B\boldsymbol{\beta} \tag{2.18}$$

Equations (2.17)–(2.18) are the major formulas for the HDCA algorithm and enables us to assembly bodies A and B into an articulated body C. Thereupon, the whole multibody system hierarchic assembly based on the binary tree decomposition is achievable (see Fig. 3). The step finishes when the root node correlated

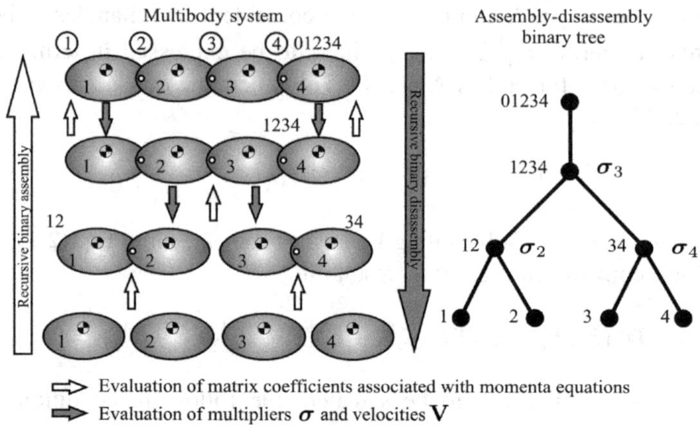

Fig. 3 Assembly-disassembly process for a multibody system

with the mechanism's (floating) base body is reached. At this point, the mechanism's base body connection conditions can be applied.

The momentum conservation equations for the whole compound free-floating system (see Fig. 3) in the form of Eqs. (2.9)–(2.10) simplify to the expression

$$\mathbf{V}_1^C = \xi_{10}^C \tag{2.19}$$

as the impulsive constraints forces $\mathbf{T}_1^C$ and $\mathbf{T}_2^C$ are both equal to $\mathbf{0}$. The spatial velocity $\mathbf{V}_1^C$ is related to the virtual kinematic pair, which has 6 degrees of freedom with respect to the inertial reference frame.

During the back propagation phase all other impulsive forces, absolute velocities and joint velocities can be computed (cf. Fig. 3). The values of outboard constraints impulsive forces $\mathbf{T}_1^C$ and $\mathbf{T}_2^C$ can be sent to proceeding computational nodes to compute inboard constraint impulsive forces by Eq. (2.16) and spatial velocities by Eqs. (2.5)–(2.8), consequently. Afterwards, joint velocities are determinable by projecting Eq. (2.11) onto the appropriate joint's motion subspace.

2.2 *Derivatives of Canonical Momenta*

In this section the formulas for time derivatives of the joint canonical momenta are derived. Once more, let us consider bodies A and B as presented in Fig. 2. The equations of motion for these bodies are as follows (**Q**—spatial external loads)

$$\dot{\mathbf{P}}_1^A + \dot{\mathbf{S}}_1^A \mathbf{P}_1^A = \mathbf{Q}_1^A + \mathbf{F}_1^A + \mathbf{S}_{12}^A \mathbf{F}_2^A \tag{2.20}$$

$$\dot{\mathbf{P}}_1^B + \dot{\mathbf{S}}_1^B \mathbf{P}_1^B = \mathbf{Q}_1^B + \mathbf{F}_1^B + \mathbf{S}_{12}^B \mathbf{F}_2^B \tag{2.21}$$

Matrices $\dot{\mathbf{S}}_1^A$ and $\dot{\mathbf{S}}_1^B$ contain absolute, translational velocities at handles 1. The quantities $\mathbf{F}_1$ and $\mathbf{F}_2$ are accordingly reaction forces, which have to satisfy the third Newton's law $\mathbf{F}_1^B = -\mathbf{F}_2^A$. Substituting $\mathbf{F}_1^B$ from Eq. (2.21) into Eq. (2.20) enables us to formulate the equation of motion for articulated body C as

$$\frac{d}{dt}\left(\mathbf{P}_1^A + \mathbf{S}_{12}^A \mathbf{P}_1^B\right) + \dot{\mathbf{S}}_1^A\left(\mathbf{P}_1^A + \mathbf{S}_{12}^A \mathbf{P}_1^B\right) = \left(\mathbf{Q}_1^A + \mathbf{S}_{12}^A \mathbf{Q}_1^B\right) + \mathbf{F}_1^A + \mathbf{S}_{12}^A \mathbf{S}_{12}^B \mathbf{F}_2^B, \tag{2.22}$$

which can be rewritten in a more concise form, similar to Eqs. (2.20)–(2.21)

$$\dot{\mathbf{P}}_1^C + \dot{\mathbf{S}}_1^C \mathbf{P}_1^C = \mathbf{Q}_1^C + \mathbf{F}_1^C + \mathbf{S}_{12}^C \mathbf{F}_2^C \tag{2.23}$$

Equation (2.23) describes the motion of the articulated body C consisting of A, B and interconnecting joint (cp. Fig. 2). Vectors $\mathbf{P}_1^C$ and $\mathbf{Q}_1^C$ include accumulated momenta and accumulated external loads. Using the above scheme, it would be possible to develop the whole multibody system hierarchic assembly according to the binary tree decomposition. By reaching the mechanism's base body the procedure finishes this step. The projection of Eq. (2.23) onto the joint's motion subspace allows one to designate the derivatives of canonical momenta

$$\dot{\mathbf{p}}_1 = \mathbf{H}_1^T\left(\mathbf{Q}_1^C - \dot{\mathbf{S}}_1^C \mathbf{P}_1^C\right) + \dot{\mathbf{H}}_1^T \mathbf{P}_1^C \tag{2.24}$$

The reaction forces $\mathbf{F}_1^C$, $\mathbf{F}_2^C$ are zero as they refer to the tips of free-floating manipulator. Based on the divide and conquer algorithm, it is possible to compute all derivatives of joint's momenta in the back-substitution phase. As $\mathbf{P}_1^C$ and $\mathbf{Q}_1^C$ vectors are unknown, the more efficient approach is to express the equations of motion in terms of articulated momenta $\overline{\mathbf{P}}$ and articulated external loads $\overline{\mathbf{Q}}$. Articulated momenta vectors are already computed in the previous step (see Sect. 2.1). For articulated external loads it is possible to construct assembly-disassembly procedure, which exhibits logarithmic computational cost in parallel

$$\mathbf{Q}_1^A = \overline{\mathbf{Q}_1^A} + \mathbf{S}_{12}^A \overline{\mathbf{Q}_2^A} \tag{2.25}$$

$$\mathbf{Q}_2^A = \mathbf{S}_{21}^A \overline{\mathbf{Q}_1^A} + \overline{\mathbf{Q}_2^A} \tag{2.26}$$

A closer look at Eqs. (2.3)–(2.4) enables us to transform expression (2.24)

$$\dot{\mathbf{p}}_k = \mathbf{H}_k^T \left(\overline{\mathbf{Q}_1^K} - \dot{\mathbf{S}}_1^{K} \overline{\mathbf{P}_1^K} \right) + \dot{\mathbf{H}}_k^{T} \overline{\mathbf{P}_1^K} \tag{2.27}$$

The derivative of canonical momenta related to joint k and body K can be computed directly as all necessary terms are known. Equation (2.27) is equivalent to the second set of Hamilton's canonical equations for free-floating manipulators possessing tree-like topologies. There is no need to construct system's Hamiltonian and its partial derivatives.

3 Numerical Example

To validate the HDCA algorithm presented here, simulation results from the modeling of one numerical test case are reported in the following paragraphs. The dynamics of multi-arm space robot is investigated in a simplified scenario, in which the robot is chasing and capturing a nonmoving object. As depicted in Fig. 4 the free-flying system includes three open kinematic chains, two of which are four-link manipulators, while the third one, is an additional appendage fulfilling some functions e.g. it can be a communicating antenna. The bodies in the system are interconnected by pin joints with the axis of revolution perpendicular to the plane of motion of the system. The exception from that rule is the floating base body 1 (spacecraft), whose connection with respect to the inertial reference frame can be regarded as three degrees of freedom generalized joint. It is assumed that each generic link of the system for $i = 2, \ldots, 10$ is rigid of length $l_i = 1$ m, mass $m_i = 1$ kg, and moment of inertia $J_{Ci} = 1$ kgm^2 with respect to the axis perpendicular to

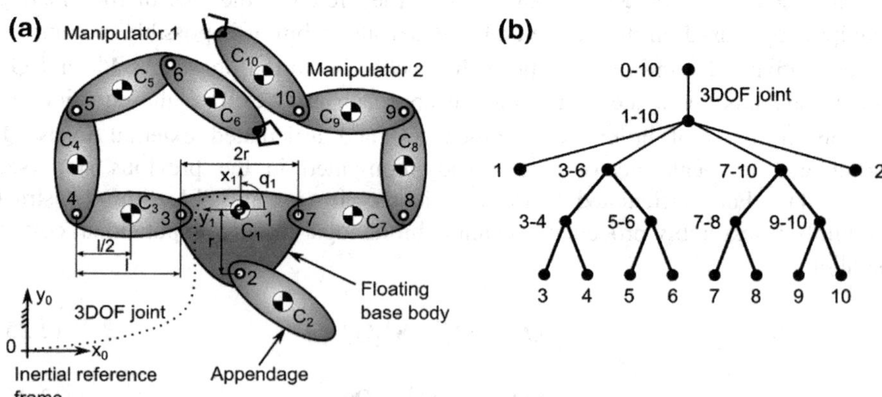

Fig. 4 A planar three-arm robotic system with floating base body. **a** The bodies and joints are numbered consecutively; characteristic lengths and body mass centers are presented. **b** The graph associated with the flow of computations is depicted. Each node in the graph represents a small portion of computations that can be performed in parallel at each level of the graph

the plane of motion and passing through the center of mass C_i. For the floating base body it is assumed the following data $r = 0.5$ m, $m_1 = 10$ kg, $J_{Ci} = 10$ kgm^2. The spacecraft is equipped with reaction jets to provide the required control forces and torques for the base body, whereas the manipulators possess actuators at joints. At the initial instant the floating base body is located at the origin of the inertial frame, i.e. $x_1^{(0)} = 0$, $y_1^{(0)} = 0$ with the angle of rotation $q_1^{(0)} = 90°$. The joint angles for the manipulator 1 and 2 are chosen in such a way that at time $t = 0$ they are spread, whereas the appendage is in the vertical bottom down position. For simplicity the velocity initial conditions for the floating base body and angular rates at joints are set to zero.

The simulation scenario assumes that the floating base body is changing its location from the initial position to the final position at point $x_1 = 8$ m, $y_1 = 6$ m. The desired orientation of the spacecraft is set to $q_1 = 45°$. At the same time two control tasks are simultaneously performed. Firstly, the appendage should maintain its bottom down orientation. Secondly, the joint angles of the two manipulators are chosen in such a way (by solving the inverse kinematics problem) to grasp the target located at point $x_t = 10$ m, $y_t = 8$ m. All desired velocities are set to zero. Since the main purpose of this investigation is to show the efficient algorithm for branched systems, it is assumed that the spacecraft control forces and torques, and the joint control torques for the manipulators are evaluated by using simple independent proportional-derivative controllers having uniformly chosen proportional and derivative gains as $k_p = 100$, $k_d = 100$. The gains are chosen experimentally to guarantee the overshoot of the response as small as possible.

The results shown in Fig. 5 are for a 5 s simulation and present the control errors for the base body orientation q_1 and joint angles $q_3, \ldots, q_6$ associated with the first manipulator. Moreover, control errors for the spacecraft position in x and y directions are depicted. On the other hand, Fig. 5 demonstrates the control signals associated with the applied independent PD controllers. It can be seen from the numerical outcome that the chosen simple control strategy ensures the stabilizing properties for the robotic system.

Figure 6 depicts a frame by frame animation of the maneuver. The results are in a good agreement with the assumed control objectives. The base body is changing its position and orientation to the desired values. The appendage is controlled well to assure the bottom down position. Also, the arms of the manipulators grasp an object appropriately.

Although the test case presented here is simple, the HDCA algorithm demonstrated in the paper can easily deal with spatial multibody robotics systems that possess tree-like topologies. An efficient HDCA recursive algorithm for evaluation of velocities and momenta of the system has the structure ideally suited for parallel computations. One can exploit this formulation to be usable in e.g. real-time control of complex multi-branched space robotic systems by reformulating its dynamics to the operational space (Bhalerao et al. 2013).

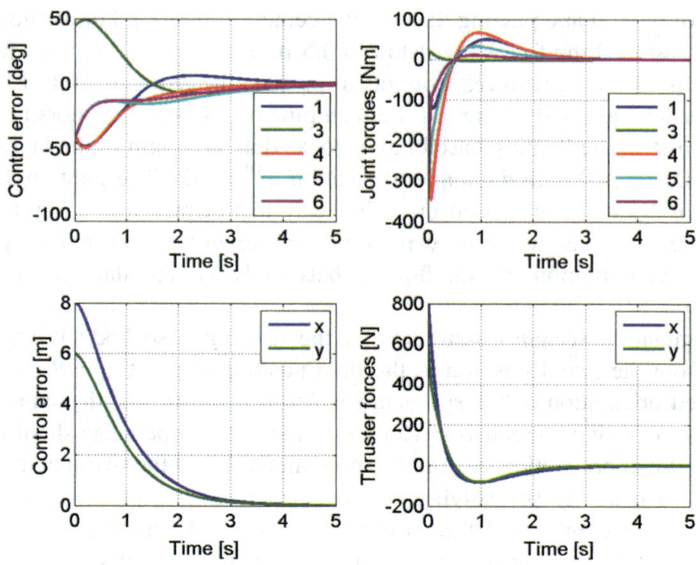

Fig. 5 The simulation results: control errors and signals for the base body and manipulator 1

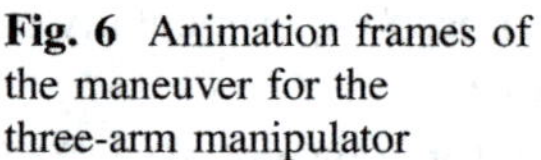

Fig. 6 Animation frames of the maneuver for the three-arm manipulator

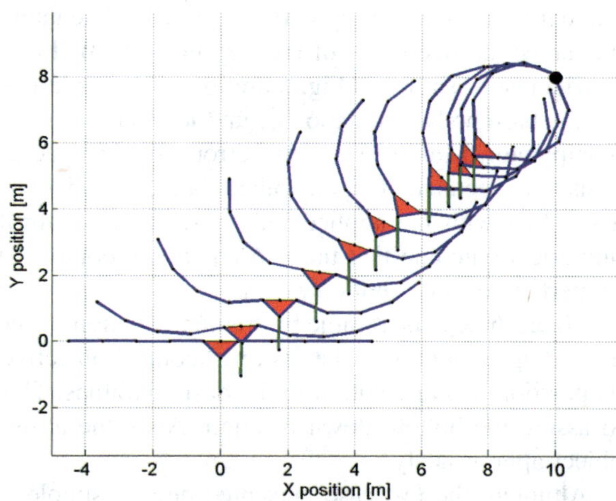

4 Discussion

4.1 Modeling Issues

In this paper the equations of motion for the free-flying multi-arm manipulator are formulated in terms of joint canonical coordinates. Herein, the Hamiltonian based parallel algorithm is presented that enables one to efficiently and robustly simulate complex spatial multi-rigid-body systems that may be subjected to independent holonomic constraints. The proposed formulation might be extended in a number of ways. One such indispensable development is the application of the HDCA algorithm for the simulation of free-flying systems with closed kinematic loops. Such situation may happen when coordinated robots mounted on spacecraft are holding an object that is supposed to be manipulated. The interaction between this object and multi-arm robotic system imposes additional constraints on the system and forms closed-kinematic chains. The generalized Hamiltonian based formulation for multi-rigid-body systems with closed-kinematic loops has been recently published in (Chadaj et al. 2017b) and it may be used in the context of free-flying systems.

The analysis of free-flying multibody systems that involve closed-loops introduces new problems of numerical and modeling nature. In analysis of such closed-loop systems the canonical coordinates and momenta are no longer independent due to loop-closure constraints. Usually, such constraints are imposed at the acceleration level for acceleration based formulations (such as Newton-Euler algorithms). In the case of the HDCA algorithm discussed here, the loop-closure constraints are imposed at the velocity level. These conditions ensure that the velocity errors are supposed to be small during the simulation. Nevertheless, numerical errors during integration (e.g. truncation errors) may introduce unbounded accumulation of position constraint errors. The problem may be solved by myriad of stabilization techniques, projective methods and criteria (Malczyk and Frączek 2012), and reduction of the dependent set of state coordinates into independent set of generalized coordinates. Other issue that might arise during the simulation is the problem of total energy conservation for a multibody system. Due to the fact that the constraints are imposed at the velocity level the HDCA algorithm indicates marginal energy drift compared to acceleration based formulations as clearly indicated in (Chadaj et al. 2017b). Obviously, the HDCA method requires some form of stabilization techniques to be involved in the formulation, especially, when long-time simulation scenarios are considered. Efficient inclusion of stabilization techniques for the Hamiltonian based parallelizable formulation is a matter of current endeavors for the authors.

When a closed-loop free-flying multi-rigid-body systems are considered it may happen that there are redundant constraints imposed on the system, i.e. conditions that can be removed without changing the kinematics of a system. The redundancy manifests itself in the inability of the HDCA algorithm to evaluate constraint force impulses that may indeed be useful in design and control applications for space robots. In this case some of the matrices in the HDCA algorithm may become permanently rank deficient. The redundant constraints should be carefully deleted from the analysis in order to proceed with the simulation. The elimination of additional constraints poses a problem of uniqueness of constraint loads at joints that is extensively discussed in Wojtyra and Frączek (2012, 2013), Pękal and Frączek (2016) together with numerical criteria for solvability of constraint loads in multi-rigid- and multi-flexible-body systems with redundant constraints. The mentioned rank deficiency may happen occasionally due to the kinematic singularities investigated here (Malczyk and Frączek 2012). This may introduce additional constraint violation errors that prevent the dynamicists from succeeding the analysis of a system.

4.2 Parallel Performance Results

The HDCA algorithm presented in this paper is the approach that enables one to efficiently simulate the dynamics of complex multi-rigid-body systems in a highly parallelizable manner. The full exploitation of parallel computing requires generation of well-balanced binary tree associated with the topology of a space system (compare Fig. 4b). The parallelization may be pursued by looking at the way the independent subassemblies are constructed when solving Hamilton's equations. First of all one may look at each level of the binary tree. Each node belonging to this set may be assigned to one thread. Thus, for instance, for leaf nodes from 3 to 10 in Fig. 4b one might exploit 8 independent threads, whereas for the nodes (subassemblies) 3–4, 5–6, 7–8, and 9–10 one might use 4 threads for calculations, etc. Obviously, this kind of mapping makes the process dynamic and fine grained, especially when kinematic chains of different lengths are attached to the spacecraft. The computational load alters as the algorithm walks up and down the binary tree in the assembly-disassembly process. Theoretically, the HDCA method has linear computational complexity $O(n)$ in terms of n—number of bodies when the calculations are performed sequentially. On the other hand, the HDCA exhibits logarithmic numerical cost $O(\log_2 n)$ when parallel computing procedures are employed on n processors. The sequential and parallel efficiency of the HDCA algorithm for serial kinematic chains is presented in the recent paper of the authors (Chadaj et al. 2017a). Such open-loop chain system may be treated as an approximation of one arm of the space system attached to the spacecraft. Therefore, the performance results collected in the cited paper may closely approximate the efficiency of the HDCA method for space systems discussed herein. Let us consider a multi-rigid-link pendulum, which is released from initial configuration due to the gravity forces. One-second simulation scenarios are considered. The Runge-Kutta of fourth order integration routine is used by assuming fixed time-step 0.005 s. On the programming end, the HDCA algorithm is implemented on GeForce GTX 960 graphics processor unit with the use of CUDA C/C++ threads. The number of GPU threads is varied from 1 to 256. Table 1 presents the performance results for different number of bodies in the system n and different number parallel computing resources (threads) t used for calculations.

As indicated in Table 1 the real benefits associated with parallel implementation of the HDCA algorithm on GPU may be obtained for long chains starting from $n = 128$ bodies in the system when at least $t = 64$ threads are employed for calculations. For shorter chains it is hardly possible to find parallel computing benefits, partially, due to the fact that the overhead associated with GPU thread management and GPU-CPU communication bottleneck degrade the performance of the

Table 1 The performance results for the HDCA algorithm. The shaded entries of the table indicate parallel speedups with respect to serial implementations on one GPU thread

	t=1	*t=2*	*t=4*	*t=8*	*t=16*	*t=32*	*t=64*	*t=128*	*t=256*
n=128	6.82	6.84	6.84	6.85	6.84	6.85	5.13	4.12	–
n=256	14.70	14.72	14.74	14.75	14.76	14.76	10.23	6.83	6.26
n=512	34.13	34.13	34.20	34.23	34.22	34.23	22.20	13.18	11.12
n=1024	85.36	85.50	85.38	85.52	85.51	85.61	52.20	28.96	22.43
n=2048	237.37	236.94	237.36	237.64	237.95	237.87	131.12	73.00	51.31

algorithm. Nevertheless, parallel speedup $S = 4.63$ is reported for $n = 2048$ bodies and $t = 256$ threads exploited in the calculations, which represents a significant increase in efficiency compared to the sequential case. For tree-like topologies, the HDCA parallel performance is rather insensitive to the type of joints existing in the analyzed chain. However, the benefits of parallel computing is more effective when longer chains are attached to the spacecraft.

5 Conclusions

In this paper we proposed an efficient recursive algorithm for the dynamics of robotic systems possessing tree-like topologies. The HDCA formulation is linear when sequential calculations are considered, and achieves optimal logarithmic complexity in parallel. The approach is ideally suited for complex robotic systems, especially, when kinematic chains are long. The HDCA formulation for dealing with generalized free-flying systems with internal closed-loops within the topology and parallel performance for such systems is under current research efforts for the authors.

Acknowledgements This work has been supported by the National Science Centre under grant no. DEC-2012/07/B/ST8/03993.

References

Abiko S, Hirzinger G (2008) Computational efficient algorithms for operational space formulation of branching arms on a space robot. In: Proceedings of IEEE IROS. https://doi.org/10.1109/iros.2008.4651048

Bhalerao K, Critchley J, Oetomo D, Featherstone R, Khatib O (2013) Distributed operational space formulation of serial manipulators. J Comput Nonlinear Dyn. https://doi.org/10.1115/1.4025577

Chadaj K, Malczyk P, Frączek J (2015) Efficient parallel formulation for dynamics simulation of large articulated robotic systems. In: Proceedings of the 20th IEEE international conference on methods and models in automation and robotics, Międzyzdroje, Poland

Chadaj K, Malczyk P, Frączek J (2017a) A parallel recursive Hamiltonian algorithm for forward dynamics of serial kinematic chains. IEEE Trans Robot. https://doi.org/10.1109/TRO.2017.2654507

Chadaj K, Malczyk P, Frączek J (2017b) A parallel Hamiltonian formulation for forward dynamics of closed-loop multibody systems. Multibody Syst Dyn 39(1):51–77. https://doi.org/10.1007/s11044-016-9531-x

Chang K, Khatib O (2000) Operational space dynamics: efficient algorithms for modeling and control of branching mechanisms. In: Proceedings of IEEE ICRA. https://doi.org/10.1109/robot.2000.844156

Dubowsky S, Papadopoulos E (1993) The kinematics, dynamics, and control of free-flying and free-floating space robotic systems. IEEE T Robot Autom 9(5):531–543

Featherstone R (1983) The calculation of robot dynamics using articulated-body inertias. Int J Robot Res 2:13–30

Featherstone R (1999) A divide-and-conquer articulated body algorithm for parallel O (log n) calculation of rigid body dynamics. Part 1: basic algorithm. Int J Robot Res 18:867–875

Jain A, Rodriguez G (1995) Base-invariant symmetric dynamics of free-flying manipulators. IEEE T Robot Autom 11(4):585–597

Laflin J, Anderson K, Khan I, Poursina M (2014) Advances in the application of the divide-and-conquer algorithm to multibody system dynamics. J Comput Nonlinear Dyn 9(4). https://doi.org/10.1115/1.4026072

Lankarani H, Nikravesh P (1988) Application of the canonical equations of motion in problems of constrained multibody systems with intermittent motion. Adv Des Autom 1:417–423

Malczyk P, Frączek J (2008) Cluster computing of mechanisms dynamics using recursive formulation. Multibody Syst Dyn 20(2):177–196

Malczyk P, Frączek J (2012) A divide and conquer algorithm for constrained multibody system dynamics based on augmented Lagrangian method with projections-based error correction. Nonlinear Dyn 70(1):871–889. https://doi.org/10.1007/s11071_012_0503_2

Malczyk P, Frączek J (2015) Molecular dynamics simulation of simple polymer chain formation using divide and conquer algorithm based on the augmented Lagrangian method. J Multi-body Dyn 229(2):116–131

Mukherjee R, Malczyk P (2013a) Efficient approach for constraint enforcement in constrained multibody system dynamics. In: Proceedings of the ASME 2013 IDETC/CIE conference on multibody systems, nonlinear dynamics, and control, Portland, USA

Mukherjee R, Malczyk P (2013b) Parallel algorithm for modeling multi-rigid body system dynamics with nonholonomic constraints. In: Proceedings of the ASME 2013 IDETC/CIE conference on multibody systems, nonlinear dynamics, and control, Portland, USA

Naudet J et al (2003) Forward dynamics of open-loop multibody mechanisms using an efficient recursive algorithm based on canonical momenta. Multibody Syst Dyn 10(1):45–59

Papadopoulos E, Dubowsky S (1991) On the nature of control algorithms for free-floating space manipulators. IEEE Trans Robot Autom 7(6):750–758

Pękal M, Frączek J (2016) Comparison of selected formulations for multibody system dynamics with redundant constraints. Arch Mech Eng LXIII(1):93–112

Umetani Y, Yoshida K (1989) Resolved motion rate control of space manipulators with generalized Jacobian matrix. IEEE Trans Robot Autom 5(3):303–314

Vafa Z, Dubowsky S (1990) The kinematics and dynamics of space manipulators: the virtual manipulator approach. Int J Robot Res 9(4):3–21

Wojtyra M, Frączek J (2012) Joint reactions in rigid or flexible body mechanisms with redundant constraints. Bull Pol Acad Sci-Tech Sci 60(3):617–626

Wojtyra M, Frączek J (2013) Comparison of selected methods of handling redundant constraints in multibody systems simulations. J Comput Nonlinear Dyn 8(2):1–9

Yamane K, Nakamura Y (2009) Comparative study on serial and parallel forward dynamics algorithms for kinematic chains. Int J Robot Res 28(5):622–629

Yokokohji Y, Toyoshima T, Yoshikawa T (1993) Efficient computational algorithms for trajectory control of free-flying space robots with multiple arms. IEEE T Robot Autom 9(5):571–580

Nonlinear Model Predictive Control (NMPC) for Free-Floating Space Manipulator

Tomasz Rybus, Karol Seweryn and Jurek Z. Sąsiadek

1 Introduction

Manipulators began to be widely used in orbital operations in 1980s since the introduction of the Space Shuttle (Jenkins 2001). The Shuttle Remote Manipulator System (SRMS) was used to deploy, maneuver and capture payloads. This manipulator assisted astronauts during first on-orbit servicing of malfunctioned satellite during STS-41C mission in 1984 (McMahan and Neal 1984). SRMS was then used during several other servicing missions (including servicing missions to the Hubble space telescope) and during construction of the International Space Station (ISS). Another manipulator, the Mobile Servicing System (MSS), is mounted on ISS and plays a key role in station assembly and maintenance (Stieber et al. 1999). Apart from assisting astronauts and transporting payloads MSS is used for capturing unmanned resupply vehicles such as Dragon. Other manipulators are also used on ISS and were used on the Space Shuttle (e.g., Nagatomo et al. 1998). Apart from several experiments that were conducted (e.g., Hirzinger et al. 1993) manipulators on the Space Shuttle and on ISS were operated by astronauts. In case of proposed unmanned on-orbit servicing (OOS) missions (e.g., Yasaka and Ashford 1996; Xu et al. 2010) or Active Debris Removal (ADR) missions (e.g., Lampariello et al. 2013) required level of autonomy is much higher. Moreover, manipulators mentioned above are mounted on large and heavy platform (Space Shuttle or ISS), thus influence of their motion on the state of the platform is not very significant. Motions of the manipulator mounted on relatively small spacecraft can significantly influence position and orientation of this spacecraft. Thus, as opera-

T. Rybus (✉) · K. Seweryn
Space Research Centre of the Polish Academy of Sciences,
Bartycka 18a str., 00-716 Warsaw, Poland
e-mail: trybus@cbk.waw.pl

J. Z. Sąsiadek
Carleton University, Ottawa, ON K1S 5B6, Canada

J. Sasiadek, *Aerospace Robotics III*, GeoPlanet: Earth and Planetary Sciences,
https://doi.org/10.1007/978-3-319-94517-0_2

tions of manipulator (e.g., capture of target object) are usually defined in the Cartesian inertial reference frame, control of such system is much more challenging. Reaction torques and reaction forces induced by the motion of the manipulator must either be fully compensated by spacecraft Attitude and Orbit Control Systems (AOCS) or AOCS must be switched off during the capture manoeuvre and free-floating nature (Dubowsky and Papadopoulos 1993) of the system must be taken into account.

Up to now certain technologies needed for unmanned OOS or ADR missions were successfully demonstrated on-orbit. The capture of a target satellite by a chaser spacecraft equipped with a manipulator was demonstrated in 1997 during ETS-VII mission (Oda 2000) and in 2007 by Orbital Express mission (Ogilvie et al. 2008). But, although various technologies necessary for autonomous capture were tested in space, fully automatic capture by a manipulator of an uncontrolled object is yet to be performed. New technology demonstration missions are currently planned, e.g., DEOS (Reintsema et al. 2010), e.Deorbit (Hausmann et al. 2015).

During orbital capture manoeuvre operations of the control system for autonomous manipulator-equipped spacecraft can be divided into two stages: (i) trajectory planning stage and (ii) control stage, in which control system must ensure realization of selected trajectory. Trajectory planning can be performed when spacecraft is waiting in a safe point close to the target object, while controller responsible for realization of the trajectory must work in real time. Several different methods that take into account free-floating nature of the system were proposed for trajectory planning, e.g., optimization techniques (Seweryn and Banaszkiewicz 2008), Rapidly Exploring Random Trees algorithm (Rybus and Seweryn 2015), or selection of simple trajectories based on Bézier curves for singularity avoidance (Rybus et al. 2013a). In this study, we focus on the second stage, i.e., control of the manipulator following the trajectory. From a variety of control techniques proposed for such systems particularly worth mentioning are resolved rate and acceleration control based on the Generalized Jacobian Matrix (GJM) introduced by Umetani and Yoshida (1989) and control schemes that incorporate feedback linearization (e.g., Aghili 2009; Barciński et al. 2013). Review of control techniques can be found in Flores-Abad et al. (2014).

In this paper we explore the possibility of using Model Predictive Control (MPC) for controlling manipulator mounted on a free-floating spacecraft (it is assumed that AOCS is switched off during the capture maneuver and during other operations that require the use of robotic arm). MPC control technique is based on a model of the system that is used to obtain the control signal by minimizing an objective function. With the model state of the system at future time instants (horizon) is predicted and receding strategy is employed (at each instant the horizon is displaced towards the future, which involves the application of the first control signal of the sequence calculated at each step).

Predictive control was introduced in chemical plants and oil refineries in 1980s. In recent years predictive control has been successfully used in a variety of robotic applications, including control of autonomous mobile robots (e.g., Kim et al. 2004) and control of manipulators (e.g., Özsoy and Kazan 1993; Gautier 2000; Becerra et al. 2005). Control of manipulators is especially difficult due to the high nonlinearity of such system. Proposed solutions for predictive control of manipulators include linearization using the feedback of the inverse dynamics of the manipulator (Poignet and Gautier 2000; Torres et al. 2001) or linearization of the system at each time step (Valle et al. 2002). The case of a space manipulator with free-floating base is yet more challenging, as state of the system must include position and orientation of the spacecraft. Moreover, such system is nonholonomic (example application of MPC for system with nonholonomic constraints can be found in Hazry and Sugisaka 2006). The idea of using MPC for controlling space manipulator during capture maneuver is not new, but in previous studies, base of the manipulator is assumed to be fixed, i.e., influence of the manipulator motion on the state of the spacecraft is ignored (e.g., McCourt and De Silva 2006). In our approach we take into account free-floating nature of the system and we base our controller on a nonlinear model (Seweryn and Banaszkiewicz 2008).

The paper is organized as follows. Equations describing dynamics of a free-floating spacecraft-manipulator system are presented in Sect. 2. The Nonlinear Model Predictive Control (NMPC) and its application for the spacecraft-manipulator system is presented in Sect. 3, while results of exemplary numerical simulation are shown in Sect. 4. The paper concludes with summary (Sect. 5).

2 Dynamics of the Spacecraft-Manipulator System

In this section we present dynamic equations for the general case of a spatial n-DOF manipulator mounted on a spacecraft. Coordinate systems and geometrical parameters of the considered spacecraft-manipulator system are shown in Fig. 1. All equations presented herein are expressed in the inertial reference frame CS_{ine}.

We choose the following generalized coordinates to describe the system:

$$\mathbf{q}_p = [\mathbf{r}_s \quad \mathbf{\Theta}_s \quad \mathbf{\theta}]^T \tag{1}$$

where $\mathbf{r}_s$ denotes the position of the spacecraft mass center, $\mathbf{\Theta}_s$ is the orientation of the manipulator-equipped spacecraft (Euler angles), while $\mathbf{\theta}$ is the n-dimensional

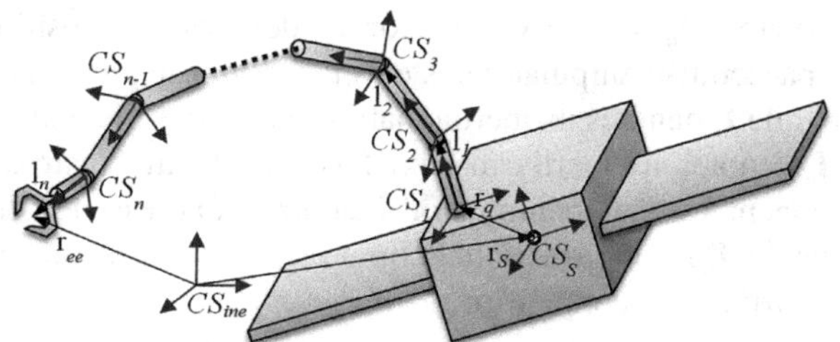

Fig. 1 Schematic view of the spacecraft-manipulator system

vector containing positions of the manipulator joints. Derivation of equations presented in this section can be found in Seweryn and Banaszkiewicz (2008). The generalized equations of motion can be presented as:

$$\mathbf{M}(\mathbf{q}_p)\ddot{\mathbf{q}}_p + \mathbf{C}(\dot{\mathbf{q}}_p, \mathbf{q}_p)\dot{\mathbf{q}}_p = \mathbf{Q} \tag{2}$$

where $\mathbf{M}$ denotes the mass matrix, $\mathbf{C}$ denotes the Coriolis Matrix and $\mathbf{Q}$ is the vector of generalized forces. No potential forces are included in Eq. (2) as the orbiting system is in the state of free fall. The mass matrix, $\mathbf{M}$, can be expressed as:

$$\mathbf{M}(\mathbf{q}_p) = \begin{bmatrix} \mathbf{A} & \mathbf{B} & \mathbf{D} \\ \mathbf{B}^T & \mathbf{E} & \mathbf{F} \\ \mathbf{D}^T & \mathbf{F}^T & \mathbf{N} \end{bmatrix} \tag{3}$$

where the submatrices $\mathbf{A}$, $\mathbf{B}$, $\mathbf{D}$, $\mathbf{E}$, $\mathbf{F}$ and $\mathbf{N}$ are defined as:

$$\mathbf{A} = \left(m_s + \sum_{i=1}^{n} m_i\right)\mathbf{I} \tag{4}$$

$$\mathbf{B} = \left(m_s + \sum_{i=1}^{n} m_i\right)\tilde{\mathbf{r}}_{s\_q} \tag{5}$$

$$\mathbf{D} = \sum_{i=1}^{n} m_i \mathbf{J}_{Ti} \tag{6}$$

$$\mathbf{E} = \mathbf{I}_s + \sum_{i=1}^{n} \left(\mathbf{I}_i + m_i \tilde{\mathbf{r}}_{i\_s}^T \tilde{\mathbf{r}}_{i\_s}\right) \tag{7}$$

$$\mathbf{F} = \sum_{i=1}^{n} \left(\mathbf{I}_i \mathbf{J}_{Ri} + m_i \tilde{\mathbf{r}}_{i\_s} \mathbf{J}_{Ti}\right) \tag{8}$$

$$\mathbf{N} = \sum_{i=1}^{n} \left(\mathbf{J}_{Ri}^T \mathbf{I}_i \mathbf{J}_{Ri} + m_i \mathbf{J}_{Ti}^T \mathbf{J}_{Ti}\right) \tag{9}$$

where $\mathbf{r}_{s\_q} = \mathbf{r}_g - \mathbf{r}_s$, (vector $\mathbf{r}_g$ denotes the position of the mass centre of the spacecraft-manipulator system) $\mathbf{r}_{i\_s} = \mathbf{r}_i - \mathbf{r}_s$, m_s denotes the mass of the spacecraft, while $\mathbf{I}_s$ denotes its inertia matrix, m_i denotes the mass of ith manipulator link, while $\mathbf{I}_i$ denotes its inertia matrix, $\mathbf{I}$ is the identity matrix, $\mathbf{J}_{Ti}$ is the translational component of the manipulator Jacobian (expressed in the inertial reference frame), while $\mathbf{J}_{Ri}$ is the rotational component of this Jacobian. Components of the Coriolis matrix, $\mathbf{C}$, are equal to:

$$\mathbf{C}_{ij} = \sum_{k=1}^{n} \left(\frac{\mathrm{d}}{\mathrm{d}(q_p)_k} m_{ij} - \frac{1}{2} \frac{\mathrm{d}}{\mathrm{d}(q_p)_i} m_{jk} \right) \tag{10}$$

where $m_{ij} \in \mathbf{M}(\mathbf{q}_p)$ and $i, j, k = 1 \ldots n$. In Eq. (10) $(q_p)_k$ denotes k-th component of the generalized coordinates vector, while $(q_p)_i$ denotes its i-th component. Vector of generalized forces $\mathbf{Q}$ in Eq. (2) can be expressed as:

$$\mathbf{Q} = [\mathbf{F}_s \quad \mathbf{H}_s \quad \mathbf{u}]^T \tag{11}$$

where $\mathbf{F}_s$ and $\mathbf{H}_s$ are forces and torques acting on the manipulator-equipped spacecraft, while $\mathbf{u}$ denotes the vector composed of driving torques applied in manipulator joints. In this study we consider free-floating spacecraft-manipulator system with zero momentum and angular momentum. Thus, we assume that there are no external forces acting on the system and: $\mathbf{F}_s = \mathbf{0}$, $\mathbf{H}_s = \mathbf{0}$. However, as shown in Seweryn and Banaszkiewicz (2008), general formulation presented up to this point can be used for the case with non-zero momentum and angular momentum.

3 Nonlinear Model Predictive Control (NMPC)

After the trajectory planning stage, the role of the control system is to ensure realization of the selected reference trajectory defined either in the configuration space or as the end-effector (EE) position in the Cartesian space. Work of the controller may be supplemented by reference control signal computed for the reference trajectory with Eq. (2). Thus, total control signal for the reference trajectory defined in the configuration space can be expressed as:

$$\mathbf{u} = \mathbf{u}_{ref} + \mathbf{u}_{contr}(\mathbf{e}_\theta, \mathbf{e}_{\dot{\theta}}) \tag{12}$$

while for the reference trajectory defined in the Cartesian space as:

$$\mathbf{u} = \mathbf{u}_{ref} + \mathbf{u}_{contr}(\mathbf{e}_p, \mathbf{e}_v) \tag{13}$$

where $\mathbf{u}_{ref}$ is an optional reference control signal computed during the trajectory planning stage ($\mathbf{u}_{ref}$ can be treated as a feedforward term and used in case of an open-loop control). When knowledge of system parameters is not perfect and when there are disturbances additional signal generated by the controller, $\mathbf{u}_{contr}$, is required to ensure realization of the reference trajectory. In case of reference trajectory defined in the configuration space this control signal $\mathbf{u}_{contr}$ should depend on error of joint positions and velocities, while for reference trajectory defined in the Cartesian space it should depend on error of EE position. Errors are defined as:

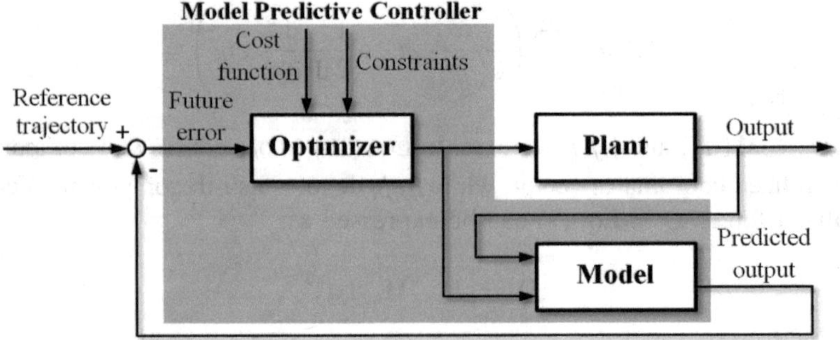

Fig. 2 Structure of predictive controller

$$\mathbf{e}_{\boldsymbol{\theta}} = \boldsymbol{\theta} - \boldsymbol{\theta}_{ref} \tag{14}$$

$$\mathbf{e}_{\dot{\boldsymbol{\theta}}} = \dot{\boldsymbol{\theta}} - \dot{\boldsymbol{\theta}}_{ref} \tag{15}$$

$$\mathbf{e}_p = \mathbf{r}_{ee} - (\mathbf{r}_{ee})_{ref} \tag{16}$$

$$\mathbf{e}_v = \mathbf{v}_{ee} - (\mathbf{v}_{ee})_{ref} \tag{17}$$

where $\mathbf{r}_{\mathrm{ee}}$ is the EE position in CS_{ine}, while $\mathbf{v}_{\mathrm{ee}}$ is the EE velocity. Subscript *ref* denoted reference trajectory obtained during trajectory planning stage.

For control of the manipulator mounted on a free-floating spacecraft we propose Nonlinear Model Predictive Control (NMPC). NMPC is an optimization based method for the feedback control of nonlinear systems (Camacho and Bordons 2007). Schematic view of the controller structure is presented in Fig. 2.

NMPC generates control torques $\mathbf{u}_{contr}$ that are applied in manipulator joints to ensure realization of the reference trajectory (remaining components of $\mathbf{Q}$ vector are zero—we are considering the free-floating case). In our approach we assume that there is no reference control torque, $\mathbf{u}_{ref}$. Thus, the proposed NMPC is responsible for ensuring realization of the reference trajectory without the feedforward term. In the NMPC controller we choose the following state vector:

$$\mathbf{x} = \begin{bmatrix} \mathbf{v}_s & \boldsymbol{\omega}_s & \dot{\boldsymbol{\theta}} & \mathbf{r}_s & \boldsymbol{\Theta}_s & \boldsymbol{\theta} \end{bmatrix}^T \tag{18}$$

Equation (2) is transformed to the following form:

$$\ddot{\mathbf{q}}_p = \mathbf{M}^{-1}(\mathbf{Q} - \mathbf{C}\dot{\mathbf{q}}_p) \tag{19}$$

and we obtain the following equations of the plant:

$$\dot{\mathbf{x}} = \begin{bmatrix} \mathbf{M}^{-1}\left(\mathbf{Q} - \mathbf{C}\begin{bmatrix} \mathbf{v}_s & \boldsymbol{\omega}_s & \dot{\boldsymbol{\theta}} \end{bmatrix}^T\right) \\ \begin{bmatrix} \mathbf{v}_s & \boldsymbol{\omega}_s & \dot{\boldsymbol{\theta}} \end{bmatrix}^T \end{bmatrix} \tag{20}$$

Equation (20) is used for simulation of system behaviour and as a plant model for NMPC. For the case of reference trajectory defined in the configuration space components of state vector $\mathbf{x}$ can be used directly in the controller, while for the reference trajectory defined in the Cartesian space position of the EE must be computed in every time step. Thus, state vector is extended to include EE position:

$$\mathbf{x} = \begin{bmatrix} \mathbf{v}_s & \boldsymbol{\omega}_s & \dot{\boldsymbol{\theta}} & \mathbf{r}_s & \boldsymbol{\Theta}_s & \boldsymbol{\theta} & \mathbf{r}_{ee} \end{bmatrix}^T \tag{21}$$

Numerical simulations were performed in Matlab environment. ACADO Toolkit implementation of the proposed NMPC control system was used. ACADO Toolkit is an algorithm collection and software environment for automatic control and dynamic optimization (Houska et al. 2011). Minimization of Least Squares Term was selected as the optimization objective on NMPC (Houska et al. 2013). However, in this approach it is not guaranteed that the global minimum will be found.

4 Results

4.1 Parameters of the System and NMPC Controller

In order to explore the possibility of using NMPC for free-floating spacecraft-manipulator system we performed simulations for a simplified example of a planar system consisting of a spacecraft with 2 DoF manipulator. Such simplification should be acceptable for our initial investigations and is common in analysis of control systems for space robotics. Equations presented in Sects. 2 and 3 are still valid for the planar case. Simulations presented in this section were performed with parameters of the planar spacecraft-manipulator system from the air-bearing microgravity simulator operated at the Space Research Centre PAS (Rybus et al. 2013b). Geometrical and mass properties of the system are shown in Table 1.

In the simplified planar case state vector $\mathbf{x}$ has 10 components:

$$\mathbf{x} = \begin{bmatrix} (v_s)_x & (v_s)_y & \omega_s & \dot{\theta}_1 & \dot{\theta}_2 & (r_s)_x & (r_s)_y & \Theta_s & \theta_1 & \theta_2 \end{bmatrix}^T \tag{22}$$

The following LSQ weighting matrix was chosen for the reference trajectory defined in the configuration space:

Table 1 Properties of the planar spacecraft-manipulator system

	Parameter	Value
1	Spacecraft mass	12.9 kg
2	Spacecraft moment of inertia	0.208 kg m^2
3	Dist. from spacecraft mass centre to manipulator mounting	[0.327 m 0]
4	Link 1 mass	4.5 kg
5	Link 1 moment of inertia	0.32 kg m^2
6	Link 1 length	0.62 m
7	Link 2 mass	1.5 kg
8	Link 2 moment of inertia	0.049 kg m^2
9	Link 2 length	0.6 m

$$\mathbf{Q}_{conf} = eye([0 \quad 0 \quad 0 \quad 0 \quad 0 \quad 0 \quad 0 \quad 0 \quad 10 \quad 10]) \tag{23}$$

As a result NMPC will optimize trajectory to minimize errors in joint positions, while other components of state vector are not taken into account. For the case of reference trajectory defined in the Cartesian space state vector $\mathbf{x}$ was extended to include EE position:

$$\mathbf{x} = \begin{bmatrix} (v_s)_x & (v_s)_y & \omega_s & \dot{\theta}_1 & \dot{\theta}_2 & (r_s)_x & (r_s)_y & \Theta_s & \theta_1 & \theta_2 & (r_{ee})_x & (r_{ee})_y \end{bmatrix}^T \tag{24}$$

and the following weighting LSQ matrix was chosen:

$$\mathbf{Q}_{cart} = eye([0 \quad 0 \quad 0 \quad 0 \quad 0 \quad 0 \quad 0 \quad 0 \quad 0 \quad 0 \quad 10 \quad 10]) \tag{25}$$

NMPC control horizon was set to 0.5 s for the configuration space and 1 s for Cartesian space with 10 control intervals for both cases.

4.2 Configuration Coordinates

First simulation was performed with reference trajectory defined in the configuration coordinates (with step changes of reference values). This reference trajectory and positions of manipulator joints obtained from simulation are presented in Fig. 3. State of the spacecraft (orientation of spacecraft and position of its center of mass in respect to the initial position) is shown in Fig. 4. In Fig. 3 it can be seen that positions of manipulator joints begin to change before the change of the reference signal. This is possible because the controller takes into account future reference signals—the resulting error in reference trajectory following is minimized. In Fig. 4 free-floating nature of the system is clearly visible.

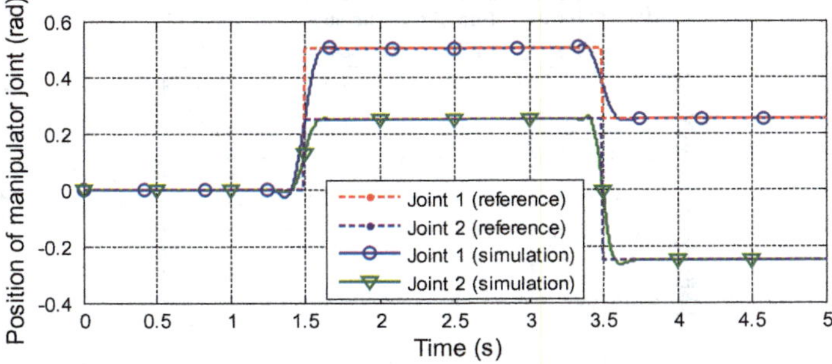

Fig. 3 Position of manipulator joints (reference vs. simulation with NMPC)

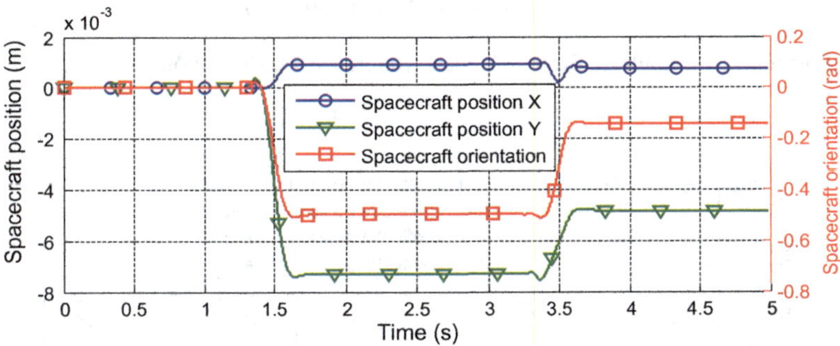

Fig. 4 Orientation of spacecraft and position of its center of mass in respect to the initial position during realization of trajectory defined in the configuration space

4.3 Cartesian Coordinates

Square was selected as a reference trajectory of the EE defined in the Cartesian space (such square trajectory is often used for evaluation of control methods for manipulators, e.g., Ulrich et al. 2012). This reference trajectory and positions of the EE obtained from simulation are presented in Fig. 5 in XY plane and in Fig. 6 as a function of time. In Fig. 7 difference between the reference EE position and EE position obtained from simulation is shown. Positions of manipulator joints during realization of the end-effector trajectory defined in the Cartesian space are presented in Fig. 8, while control torques applied at manipulator joints are shown in Fig. 9. The controller is able to ensure proper realization of the trajectory and errors are relatively small. Deviations from the reference trajectory are seen near the corners of the square. As in the case with reference trajectory defined in the configuration coordinates, motions of the manipulator influence state of the spacecraft. Comparison between the results obtained with NMPC and two other control algorithms (Modified Simple Adaptive Control and algorithm based on the Dynamic Jacobian inverse) can be found in Rybus et al. (2017).

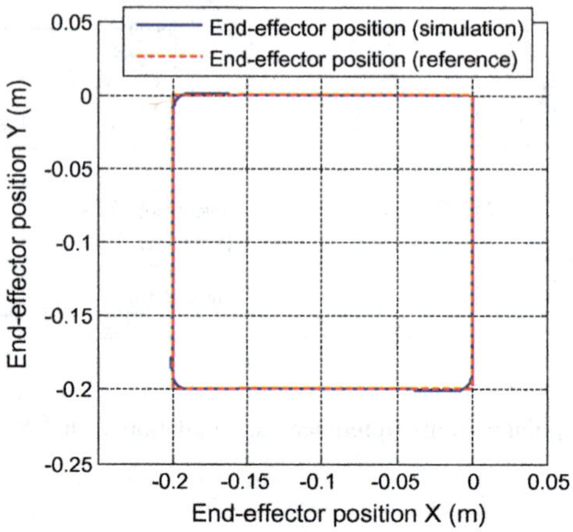

Fig. 5 Position of manipulator EE in XY plane (reference vs. simulation)

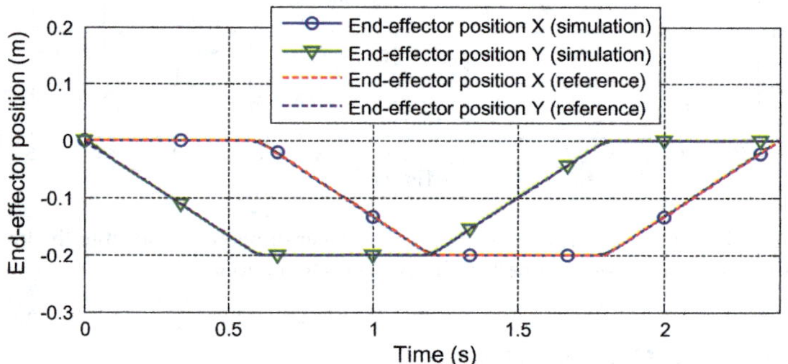

Fig. 6 Position of manipulator EE as a function of time (reference vs. simulation)

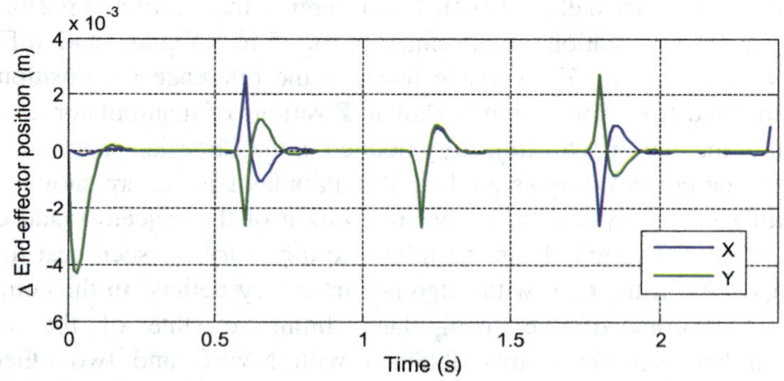

Fig. 7 Difference between the reference EE position and EE position obtained from simulation

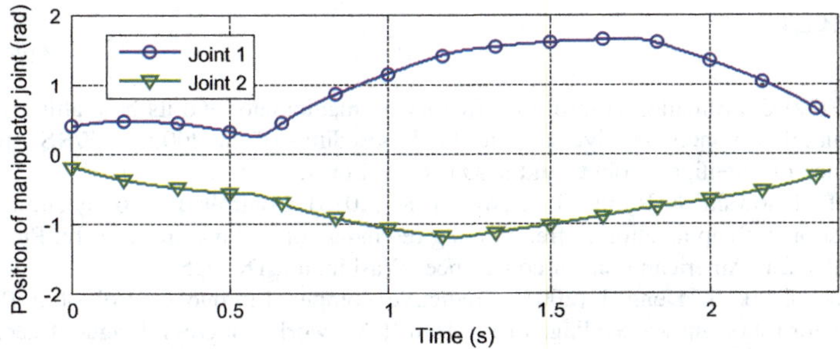

Fig. 8 Positions of manipulator joints (simulation) during realization of the EE trajectory defined in the Cartesian space

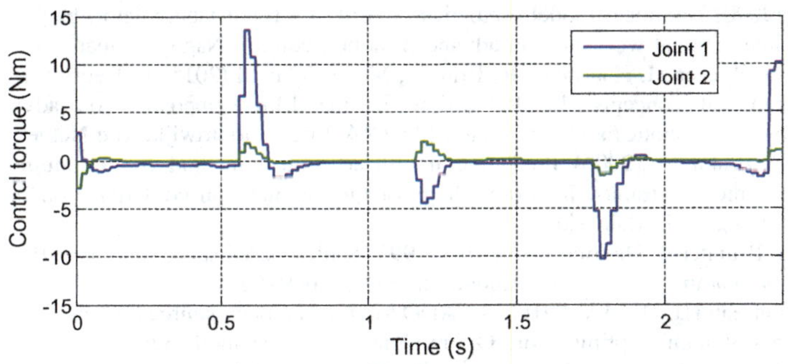

Fig. 9 Control torques applied at manipulator joints (obtained from NMPC)

5 Summary

In this paper possibility of applying Nonlinear Model Predictive Control (NMPC) to free-floating spacecraft-manipulator system was explored. Numerical simulations were performed with implementation of the proposed control system. These simulations were done for the planar case with 2 DOF manipulator. First simulation was performed with reference trajectory defined in the configuration coordinates, while the second simulation was performed with reference trajectory of the end-effector defined in the Cartesian space. Both simulations showed effectiveness of the NMPC. Errors in trajectory realization are small and advantages of the proposed approach are visible (controller takes into account future reference signals and can initiate motion before change of the reference signal). Future work will include comparison of NMPC with other control methods and experiments on the planar air-bearing microgravity simulator at the Space Research Centre PAS. Stability of the proposed method also needs to be proved.

References

Aghili F (2009) Coordination control of a free-flying manipulator and its base attitude to capture and detumble a noncooperative satellite. In: Proceedings of the 2009 IEEE/RSJ international conference on intelligent robots and systems, St. Louis, USA

Barciński T, Lisowski J, Rybus T, Seweryn K (2013) Controlled zero dynamics feedback linearization with application to free-floating redundant orbital manipulator. In: Proceedings of the 2013 IEEE American control conference, Washington DC, USA

Becerra VM, Cook S, Deng J (2005) Predictive computed-torque control of a PUMA 560 manipulator robot. In: Proceedings of the 16th IFAC world congress, Prague, Czech Republic

Camacho EF, Bordons C (2007) Model predictive control, 2nd edn. Springer, London

Dubowsky S, Papadopoulos E (1993) The kinematics, dynamics, and control of free-flying and free-floating space robotic systems. IEEE Trans Robot Autom 9(5):531–543

Flores-Abad A, Ma O, Pham K, Ulrich S (2014) A review of space robotics technologies for on-orbit servicing. Prog Aerosp Sci 68:1–26

Gautier M (2000) Nonlinear model predictive control of a robot manipulator. In: Proceedings of the 6th international workshop on advanced motion control, Nagoya, Japan

Hausmann G, Wieser M, Haarmann R, Brito A, Meyer JC et al (2015) E. Deorbit mission: OHB debris removal concepts. In: Proceeding of the 13th symposium on advanced space technologies in robotics and automation (ASTRA'2015), Noordwijk, The Netherlands

Hazry D, Sugisaka M (2006) Predictive nonlinear control method for a mobile robot with nonholonomic constraints. In: Proceedings of the international conference on man machine systems, Langkawi, Malaysia

Hirzinger G, Brunner B, Dietrich J, Heindl J (1993) Sensor-based space robotics—ROTEX and its telerobotic features. IEEE Trans Robot Autom 9(5):649–663

Houska B, Ferreau HJ, Diehl M (2011) ACADO toolkit—an open-source framework for automatic control and dynamic optimization. Optim Control Appl Methods 32(3):298–312

Houska B, Ferreau HJ, Vukov M, Quirynen R (2013) ACADO toolkit user's manual

Jenkins DR (2001) Space shuttle: the history of the national space transportation system, The First 100 Missions, 3rd edn

Kim B, Necsulescu D, Sasiadek J (2004) Autonomous mobile robot model predictive control. Int J Control 77(16):1438–1445

Lampariello R, Gahbler P, Sommer J (2013) Analysis of a deorbiting maneuver of a large target satellite using a chaser satellite with a robot arm. In: Proceedings of the 12th symposium on advanced space technologies for robotics and automation (ASTRA'2013), Noordwijk, The Netherlands

McCourt R, De Silva CW (2006) Autonomous robotic capture of a satellite using constrained predictive control. IEEE/ASME Trans Mechatron 11(6):699–708

McMahan T, Neal V (1984) Repairing solar max: the solar maximum repair mission. National Aeronautics and Space Administration Technical Report, USA

Nagatomo M, Harada C, Ishii Y, Kasuga K, Tanaka M, Hayashi M, Uchibori Y, Imaki K, Ito M (1998) Results of the manipulator flight demonstration (MFD) flight operation. In: Proceedings of the 5th international conference on space operations (SpaceOps), Tokyo, Japan

Oda M (2000) Summary of NASDA's ETS-VII robot satellite mission. J Robot Mechatron 12(4)

Ogilvie A, Allport J, Hannah M, Lymer J (2008) Autonomous satellite servicing using the orbital express demonstration manipulator system. In: Proceedings of the 9th international symposium on artificial intelligence, robotics and automation in space (i-SAIRAS'2008), Los Angeles, USA

Özsoy C, Kazan R (1993) Cartesian base predictive control of robotic manipulators. In: Proceedings of the IEEE international symposium on industrial electronics, Budapest, Hungary

Poignet P, Gautier M (2000) Nonlinear model predictive control of a robotic manipulator. In: Proceedings of the 6th international workshop on advanced motion control, Nagoya, Japan

Reintsema D, Thaeterm J, Rathke A, Naumann W, Rank P, Sommer J (2010) DEOS—the German robotics approach to secure and de-orbit malfunctioned satellites from low earth orbits. In: Proceedings of the 10th international symposium on artificial intelligence, robotics and automation in space (i-SAIRAS'2010), Sapporo, Japan

Rybus T, Seweryn K (2015) Application of rapidly-exploring random trees (RRT) algorithm for trajectory planning of free-floating space manipulator. In: Proceedings of the 10th international workshop on robot motion and control (RoMoCo'2015), Poznań, Poland

Rybus T, Barciński T, Lisowski J, Seweryn K, Nicolau-Kukliński J et al (2013a) Experimental demonstration of singularity avoidance with trajectories based on the Bézier Curves for free-floating manipulator. In: Proceedings of the 9th international workshop on robot motion and control (RoMoCo'2013), Wąsowo, Poland

Rybus T, Barciński T, Lisowski J, Nicolau-Kukliński J, Seweryn K et al (2013b) New planar air-bearing microgravity simulator for verification of space robotics numerical simulations and control algorithms. In: Proceedings of the 12th symposium on advanced space technologies in robotics and automation (ASTRA'2013), Noordwijk, The Netherlands

Rybus T, Seweryn K, Sasiadek JZ (2017) Control system for free-floating space manipulator based on nonlinear model predictive control (nmpc). J Intell Robot Syst 85(3):491–509

Seweryn K, Banaszkiewicz M (2008) Optimization of the trajectory of a general free-flying manipulator during the rendezvous maneuver. In: Proceedings of the AIAA guidance, navigation and control conference and exhibit (AIAA-GNC'2008), Honolulu, Hawaii, USA

Stieber ME, Hunter DG, Abramovici A (1999) Overview of the mobile servicing system for the international space station. In: Proceedings of the 5th international symposium on artificial intelligence, robotics and automation in space (i-SAIRAS'1999), Noordwijk, The Netherlands

Torres S, Méndez J, Acosta L, Sigut M, Marichal GN, Moreno L (2001) A predictive control algorithm with interpolation for a robot manipulator with constraints. In: Proceedings of the 2001 IEEE international conference on control applications, Mexico City, Mexico

Ulrich S, Sasiadek J, Barkana I (2012) Modeling and direct adaptive control of a flexible-joint manipulator. J Guidance Control Dyn 35(1):25–39

Umetani Y, Yoshida K (1989) Resolved motion rate control of space manipulators with generalized Jacobian matrix. IEEE Trans Robot Autom 5(3):303–314

Valle F, Tadeo F, Alvarez T (2002) Predictive control of robotic manipulators. In: Proceedings of the 2002 IEEE international conference on control applications, Glasgow, Scotland, UK

Xu W, Liang B, Gao D, Xu Y (2010) A space robotic system used for on-orbit servicing in the geostationary orbit. In: Proceedings of the 2010 IEEE/RSJ international conference on intelligent robots and systems, Taipei, Taiwan

Yasaka T, Ashford W (1996) GSV: an approach toward space system servicing. Earth Space Rev 5 (2):9–17 [illegible] D, [illegible] (2010) DEOS – the German robotics approach to [illegible] and deorbit malfunctioned satellites from low earth orbits. In: Proceedings of the [illegible] international symposium on artificial intelligence, robotics and automation in space (i-SAIRAS 2010), Sapporo, Japan

[illegible] (2015) [illegible] of a [illegible] [illegible] (CT) [illegible] [illegible] dynamics [illegible] space [illegible]. In: Proceedings of the 10th international workshop on robotic [illegible] (2015), [illegible]

[illegible] (2013) [illegible] [illegible] In: [illegible] [illegible] experimental [illegible] based on the [illegible] free-floating [illegible]. In: Proceedings of [illegible] [illegible] (2011) [illegible]

[illegible] et al (2011) [illegible] [illegible] [illegible] for [illegible] numerical simulation and experimental [illegible]. In: Proceedings of the [illegible] on advanced space technologies in robotics and automation (ASTRA 2011), [illegible], The Netherlands

[illegible] (2013) [illegible] [illegible] space manipulator [illegible] [illegible] control [illegible]. [illegible] [illegible]

[illegible] (2008) [illegible] [illegible] trajectory of a general free-flying [illegible] [illegible]. In: Proceedings of the [illegible] [illegible] control conference (AIAA GNC 2008), Honolulu, Hawaii, USA

[illegible] (2013) [illegible] [illegible] In: [illegible] symposium on [illegible] [illegible] (i-SAIRAS 2012), [illegible], The Netherlands

[illegible] et al (2011) [illegible] [illegible]. In: Proceedings [illegible] Munich, Germany

[illegible] (2012) [illegible] [illegible] control [illegible] [illegible]

[illegible] (2010) [illegible] [illegible] motion and control of space [illegible] [illegible]

[illegible] (2012) [illegible] [illegible] [illegible] [illegible]

[illegible] (2015) [illegible] [illegible] In: [illegible] (2010) [illegible] [illegible] international [illegible] [illegible], Japan

[illegible] (2007) [illegible] [illegible] Earth Space [illegible]

Performance Control of a Spacecraft-Robotic Arm System-Desired Motion Tracking

Elżbieta Jarzębowska

1 Introduction

For fixed-based manipulators, e.g. industrial manipulators, the base is a reference frame but for a robotic arm mounted on a spacecraft, the situation is different. The spacecraft is free of external forces and torques and not actuated, and linear momentum and angular momentum are conserved. Dynamic coupling between robotic arm and the spacecraft base is present and the spacecraft base can no longer serve as the inertial reference frame. It results in a more challenging derivation of equations of motion as well as a controller design. Also, the linear momentum conservation generates a holonomic constraint on a spacecraft and the angular momentum conservation-a nonholonomic constraint. A free-floating spacecraft due to its unactuation is then a nonholonomic underactuated control system. In control setting, the underactuation is treated as a second order nonholonomic constraint, so the free-floating spacecraft subjected to desired motion constraints is a multi-constraint control system (Jarzębowska and Pilarczyk 2015; Jarzębowska and Pietrak 2014).

Motivations for undertaking this research are two-fold. Firstly, underactuated nonholonomic control systems are challenging with respect to modeling and control designs. Variety of control approaches is applied to them; see e.g. (Fantoni and Lozano 2002; Jarzębowska 2012). Many researchers strive to develop control strategies for systems with constraints, e.g. on accelerations, and use the constraint equations only to transform them into control models, see e.g. (Hervas and Reyhanoglu 2013). Their main drawbacks are that real control inputs to a system have to be recovered by integration. It can lead to control errors, and be undesirable from a control engineer point of view. Real motion and control actions take place at

E. Jarzębowska (✉)
Institute of Aeronautics and Applied Mechanics, Warsaw University of Technology, Nowowiejska 24 St, Warsaw, Poland
e-mail: elajarz@meil.pw.edu.pl

J. Sasiadek, *Aerospace Robotics III*, GeoPlanet: Earth and Planetary Sciences,
https://doi.org/10.1007/978-3-319-94517-0_3

the dynamics level where inertia properties, external force effects, e.g. friction, and disturbances, may have significant influence on a control loop. For these reasons, constrained systems control should be designed at the dynamics level. There are attempts of taking advantage of analytical dynamics methods for control design. However, these reported in the literature, design controllers which are hard for implementations since they are nonlinear functions of state variables and their derivatives, and real inputs to systems are to be recovered by integration, see e.g. (Udwadia and Wanichanon 2014).

The second motivation is a potential significance in applications of developments of new control strategies for space robots. They may provide a new insight into nonlinear control methods for future missions in space as well as in space debris removal (Castronuovo 2011).

The idea of a virtual manipulator (VM) introduced in (Vafa and Dubowski 1990) enables taking a virtual ground as an inertial fixed point when the system has no initial linear momentum. Thanks to the VM concept, some existing control algorithms can be applied to a spacecraft-robotic arm system with some minor restrictions (Papadopoulos 1990; Papadopoulos and Dubowsky 1991a, b). The VM is just an abstract manipulator to simplify dynamic modeling but the other concept is a dynamically equivalent manipulator (DEM), which can be built for realistic experiments in the laboratory (Liang et al. 1996, 1998).

Taking advantage of the VM concept, the paper presents a model-based controller design for a robotic arm mounted on spacecraft. The arm is to perform a desired motion or move with a desired velocity or acceleration. The desired motion may be specified then by constraint equations. The constrained dynamics and control dynamics are developed based upon one multi-purpose modeling framework, which constitutes a basis for a development of an advanced control platform architecture. The control platform is a fusion of modern dynamics modeling, control algorithms and embedded controllers. Examples of desired motion tracking illustrate the theoretical development in modeling and the control platform applications.

The paper is organized as follows. In Sect. 2 a spacecraft constrained dynamics is developed including constraint specifications of both conservation laws and task-based constraints. Section 3 presents the multi-purpose modeling framework, which provides reference motions to the advanced control platform. Simulation studies for desired motions tracking are detailed in Sect. 4. The paper closes with conclusions and a list of references.

2 Constrained Dynamics of a Spacecraft-Robot Arm System

An analytical approach to dynamics modeling and control of constrained under-actuated systems starts from the formulation of a unified representation of constraints (Jarzębowska 2012), which is

$$B(t, q, \dot{q}, \ldots, q^{(p-1)})q^{(p)} + s(t, q, \dot{q}, \ldots, q^{(p-1)}) = 0 \tag{2.1}$$

where q is a n-dimensional state vector, B is a $(k \times n)$ dimensional matrix, p-constraint order, $n > k$, and s is a k-vector.

This formulation can include position and kinematic nonholonomic constraints, which are referred to as material; for the detailed constraint classification see, e.g. (Jarzębowska 2012). In control setting, the linear momentum conservation is a kinematic holonomic constraint so it can be presented by Eq. (2.1) for $p = 1$. The angular momentum conservation is a nonholonomic constraint for which $p = 1$. They both are referred to as conservation laws rather than constraints but they are handled in the same way as constraints when a controller is designed.

For $p \geq 1$, the constraints may be non-material and are referred to as programmed. They are imposed by a designer or a control engineer to obtain a system desired performance. When a trajectory is a program for motion, then $p = 0$. If we are interested in acceleration or jerk desired time histories, then $p = 2$ or $p = 3$.

The unified representation of constraints (2.1) can be transformed into a state-space control form. However, it is only formally equivalent to the form which nonlinear control theory actually uses. It is neither kinematic control model nor control inputs have physical interpretations of velocities. The constraint equations for $p > 2$ are referred to as higher order nonholonomic constraints (HONC) and they can be transformed to the state space representation

$$\dot{x} = f(x) + g(x)u(t). \tag{2.2}$$

as follows. Introduce $x = (x_1, \ldots, x_p)$, $x_1 = q, \dot{x}_1 = x_2, \ldots, \dot{x}_{p-1} = x_p$, then we may present (2.1) as

$$\begin{aligned} \dot{x}_1 &= x_2, \\ \dot{x}_2 &= x_3, \\ \vdots\ &\quad \vdots \\ \dot{x}_{p-1} &= x_p, \\ B(x_1, \ldots, x_p)\dot{x}_p &= -s(x_1, \ldots, x_p) \end{aligned} \tag{2.3}$$

or in a compact form

$$C_p(x)\dot{x} = b(x),$$

where C_p is a $(p - 1+ k) \times p$ matrix and b is a $(p - 1+ k)$—dimensional vector.

When $f(x)$ is a particular solution of (2.3), then $C_p(x)f(x) = b(x)$. Let $g(x)$ be a $p \times (n - k)$ full rank matrix whose column space is in the null space of $C_p(x)$, i.e. $C_p(x)g(x) = 0$. Then, the solution of (2.3) is given by Eq. (2.2) for any smooth vector $u(t)$. Also, when the number of constraint equations k is equal to the number of states n, then a program is referred to as complete. We consider programs which are partial, i.e. $n > k$.

A multi-purpose modeling framework for constrained mechanical systems is based on a latest dynamics modeling method, i.e. on the generalized programmed motion equations (GPME) method. It enables deriving motion equations for systems subjected to HONC (2.1). The GPME yield

$$\begin{aligned} &M(q)\ddot{q} + V(q,\dot{q}) + D(q) = Q(t,q,\dot{q}),\\ &B(t,q,\dot{q},\ldots,q^{(p-1)})q^{(p)} + s(t,q,\dot{q},\ldots,q^{(p-1)}) = 0, \end{aligned} \tag{2.4}$$

where q is a n-dimensional state vector, B is a $(k \times n)$ dimensional constraint matrix, p-constraint order, $n > k$, and s is a k-vector. Q is a vector of external forces, which are not controls. Equation (2.4) result in a constrained dynamics that enables planning desired motion. Also, using (2.4) for $p = 1$, a dynamic control model in a reduced state form, i.e. free of constraint reaction forces, can be developed.

The constrained dynamics (2.4) is referred to as a reference dynamics. The solutions to (2.4) are position time histories and their time derivatives satisfy the constraints. Notice that the dynamics (2.4) is free of the constraint reaction forces, which are eliminated in the derivation process. This is the fundamental advantage of (2.4) which makes them suitable for direct control applications. Lagrange equations in a reduced state form, i.e. after the multipliers elimination, are equivalent to the GPME for $p = 1$. The GPME (2.4) are also suitable for underactuated nonholonomic systems such as spacecraft.

The derivation of (2.4) can be automated and easily applied to any commercial software, e.g. Matlab; for details see (Jarzębowska 2012; Jarzębowska et al. 2017).

3 Control Platform Architecture for Desired Motions Tracking

Using Eq. (2.4) for $p = 1$, and taking into account material constraints on a system or, in the case of spacecraft, the conservation laws, a dynamic control model in a reduced state form can be generated. The separation of the material and conservation laws from programmed constraints is a key point that enables using directly analytical dynamics methods to nonlinear control design.

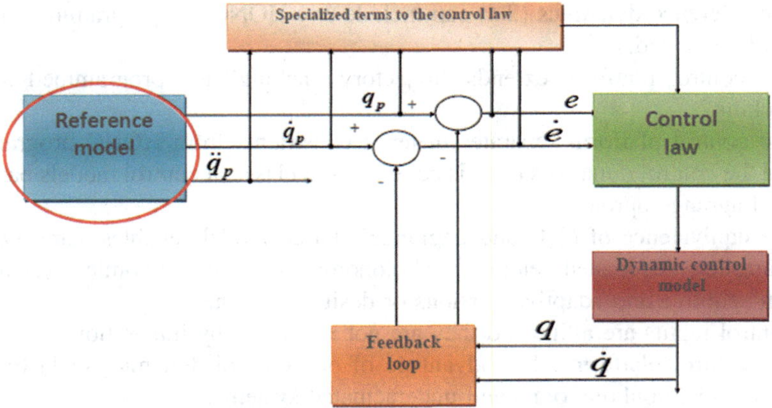

Fig. 1 Tracking control strategy architecture (Jarzębowska 2012)

The modeling framework (2.4) constitutes a basis for a development of an advanced control platform architecture, which is a fusion of modern dynamics modeling, control algorithms and embedded controllers. Architecture of a control platform is presented in Fig. 1. This architecture is modular and the resulted constrained system dynamics (2.4) enters the control loop such that it enables avoiding differentiation of control inputs. It works in the following way. Motion generated by the constrained dynamics (2.4), which includes conservation laws, material constraints and programmed constraints is referred to as a reference motion. The reference motion is a motion planner for a system desired performance. Solutions to the reference dynamics are inputs to a tracking control algorithm.

The control objective of programmed motion tracking is formulated as follows: given a reference motion specified by the constrained dynamics, design a feedback controller that can track the reference motion.

The control dynamics, to which controllers can be plugged in, is developed using the GPME including material constraints and conservation laws only, i.e. $p = 0$ or $p = 1$. In case of the spacecraft-robotic arm system modeled by joint or generalized coordinates, there are no equations of material constraints. The conservation laws are satisfied only. The control dynamics yield

$$\begin{aligned} &M(q)\ddot{q}+C(q,\dot{q})\dot{q}+D(q)=E(q)\tau,\\ &B_1(q)\dot{q}=0. \end{aligned} \tag{3.1}$$

In the overall control strategy architecture, the separation of the programmed constraints from material and conservation laws is a key point that enables using analytical dynamics methods directly to control.

Basic advantages of application of the advanced control platform can be summarized as follows:

1. The reference dynamics (2.4) can include any HONC i.e. programmed motion can be planned.
2. The control platform extends "trajectory tracking" to "programmed motion tracking".
3. The control platform separates material constraints from HONC programmed and the control dynamics (3.1) is equivalent to classical control models based on the Lagrange aproach.
4. The equivalence of (3.1) and Lagrange's based models enables using existing controllers dedicated either to holonomic or nonholonomic systems in non-adaptive and adaptive versions or design new ones.
5. Control inputs are affine and they are not recovered by integration.
6. The control platform takes advantage of one control dynamics (3.1) for both holonomic, nonholonomic and underactuated systems.
7. Modularity of the control platform enables adding new blocks, e.g. a velocity observer block.
8. The GPME method and modularity of the strategy enables using any parameters, e.g. quasi-coordinates to describe motion (Jarzębowska 2012).
9. The reference dynamic model can be generated off-line. A library of reference models that plan different tasks can be created. They all can be applied to one control dynamics of a specific system.

4 Tracking Desired Motions by a Spacecraft-Robotic Arm System-Simulation Studies

The simulation studies are based upon a simple model of a space robot equipped with a 2-link arm. It is presented in Fig. 2. The system parameters are as follows: M-mass of the base, $m_2 = m_3$-masses of the links, I_0, I_1, I_2-moments of inertia of the base and the links.

Data using for simulation studies is collected in Table 1.

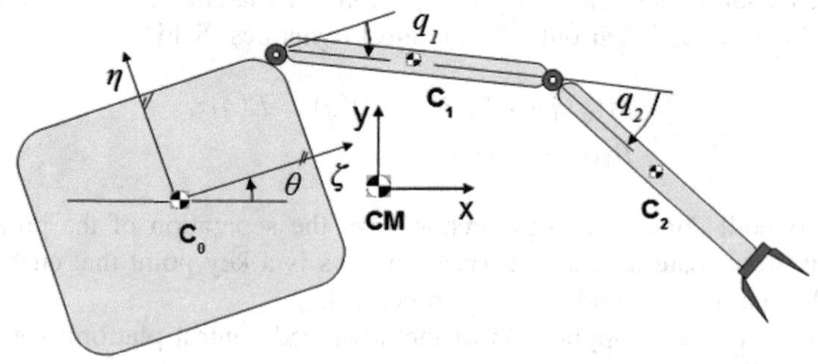

Fig. 2 The spacecraft-robotic arm system model (Jarzębowska and Pietrak 2014)

Table 1 Spacecraft model parameters

Parameter	Magnitude (unit)
Base mass M	10 (kg)
Link 1 mass m_2	1 (kg)
Link 2 mass m_3	1 (kg)
Base (cube) moment of inertia J_0	6.6 (kg m^2)
Link 1 length L_1	5 (m)
Link 2 length L_2	6 (m)
Base cube side	1 (m)
Time step Δt	0.01 (s)
Simulation time t_s	350 (s)
Coordinate	Initial condition (unit)
Base coordinate x_0	1 (m)
Base coordinate y_0	3 (m)
Base angle θ	90 (°)
Link 1 angle q_1	0 (°)
Link 2 angle q_2	170 (°)
Base velocity $\dot{x}_0$	0.01 (m/s)
Base velocity $\dot{y}_0$	−0.01 (m/s)
Base angular velocity $\dot{\theta}$	0.02 (°/s)
Link 1 angular velocity $\dot{q}_1$	−0.01 (°/s)
Link 2 angular velocity $\dot{q}_2$	0.015 (°/s)

The system linear momentum is conserved, i.e.

$$p = \sum_i m_i v_i = (M + m_1 + m_2) v_{CM} = \text{const.}, \tag{4.1}$$

what implies that

$$r_{CM} = 0. \tag{4.2}$$

The angular momentum is also conserved and assumed to be zero, i.e.

$$\begin{aligned} K = M(x_0\dot{y}_0 - \dot{x}_0 y_0) + I_0\dot{\theta} + m_1(x_1\dot{y}_1 - \dot{x}_1 y_1) + I_1\left(\dot{\theta} + \dot{q}_1\right) \\ + m_2(x_2\dot{y}_2 - \dot{x}_2 y_2) + I_2\left(\dot{\theta} + \dot{q}_1 + \dot{q}_2\right) = 0 \end{aligned} \tag{4.3}$$

what can be rewritten in a compact form as

$$K = \sum_0^2 K_i = K_0(\theta, q_1, q_2)\dot{\theta} + K_1(\theta, q_1, q_2)\dot{q}_1 + K_2(\theta, q_1, q_2)\dot{q}_2 = 0. \tag{4.4}$$

The constraints (4.1) and (4.4) are the conservation laws and can be presented by Eq. (2.1). The GPME for $p = 1$ are used to derive motion equations for the system.

The robot arm maneuvers are local, i.e. no on orbit motion is simulated. It is assumed then that the spacecraft is in a right location for the specified maneuver.

4.1 Tracking a Desired Trajectory by a 2-D Spacecraft-Robotic Arm System End-Effector

Control dynamics of a 2-D spacecraft-robotic arm system with two control inputs related to q_1 and q_2 can be presented as

$$M(q)\ddot{q} + C(q,\dot{q}) = T \tag{4.5}$$

where $T = \begin{bmatrix} 0 & 0 & 0 & \tau_2 & \tau_3 \end{bmatrix}^T = \begin{bmatrix} 0_{1\times 3} & \tau^T \end{bmatrix}^T$.

The state vector q is partitioned into unactuated q_f and actuated q_a degrees of freedom $q_f \equiv \begin{bmatrix} x_0 & y_0 & \theta \end{bmatrix}^T$, $q_a \equiv \begin{bmatrix} q_1 & q_2 \end{bmatrix}^T$.

Matrices $M(q)$ and $C(q,\dot{q})$ are partitioned such that (4.5) can be presented as

$$\begin{cases} M_{11}\ddot{q}_f + M_{12}\ddot{q}_a + C_1(q,\dot{q}) = 0, \\ M_{21}\ddot{q}_f + M_{22}\ddot{q}_a + C_2(q,\dot{q}) = \tau \end{cases} \tag{4.6}$$

Then, determining a vector of unactuated degrees of freedom from the first of equations of (4.6)

$$\ddot{q}_f = -M_{11}^{-1}[M_{12}\ddot{q}_a + C_1(q,\dot{q})] \tag{4.7}$$

and inserting into the second one yields

$$M\ddot{q}_a + D = \tau \tag{4.8}$$

with $M = M_{22} - M_{21}M_{11}^{-1}M_{12}, \quad D = C_2(q,\dot{q}) - M_{21}M_{11}^{-1}C_1(q,\dot{q})$.

Applying partial feedback linearization to (4.8) we obtain

$$\tau = Mu + D \tag{4.9}$$

Finally, control dynamics for the spacecraft-robotic arm system has the form

$$\begin{cases} \ddot{q}_a = u, \\ \ddot{q}_f = R\ddot{q}_a + H \end{cases} \tag{4.10}$$

with $R = -M_{11}^{-1}M_{12}, \quad H = -M_{11}^{-1}C_1(q,\dot{q})$.

Equation (4.10) can be presented in a standard state space control form (2.2).

To control motion of the robotic arm end-effector, a desired motion, i.e. the coordinates $(x_{Ed}(t), y_{Ed}(t))$, have to be specified by the controlled joint coordinates, i.e. $(q_{1d}(t), q_{2d}(t))$ through the geometric relations for the position vector r_E and Eq. (4.2). Notice, that for $(x_{Ed}(t), y_{Ed}(t))$ the coordinates $(q_{1d}(t), q_{2d}(t))$ are not unique for a given location of the end-effector. The base coordinates x_0, y_0 can be eliminated by Eq. (4.2) and θ is eliminated by Eq. (4.3).

The selected control algorithm is a PD with a correction, i.e.

$$\begin{aligned} u_1 &= \ddot{q}_{1d} + k_{d1}(\dot{q}_{1d} - \dot{q}_1) + k_{p1}(q_{1d} - q_1) \\ u_2 &= \ddot{q}_{2d} + k_{d2}(\dot{q}_{2d} - \dot{q}_2) + k_{p2}(q_{2d} - q_3) \end{aligned} \tag{4.11}$$

The selected trajectory is an eight shaped one, i.e.

$$x_{Ed} = 5 + 4\sin\left(2 \cdot \frac{36}{35}t - 30\right), \quad y_{Ed} = -2 + \cos\left(\frac{36}{35}t - 15\right). \tag{4.12}$$

Simulation time $t_s = 350$ s and the control gains are: $k_{p1} = k_{p2} = 2$, $k_{d1} = k_{d2} = 10$. Tracking the trajectory (4.12), tracking errors and magnitudes of joint torques are presented in Figs. 3, 4, 5, and 6.

The simulation results demonstrate that the unactuated base can translate and rotate; translation is about 10% of that of arm lengths. The control torques, at the beginning of tracking, are quite large due to initial conditions selected for end-effector E position.

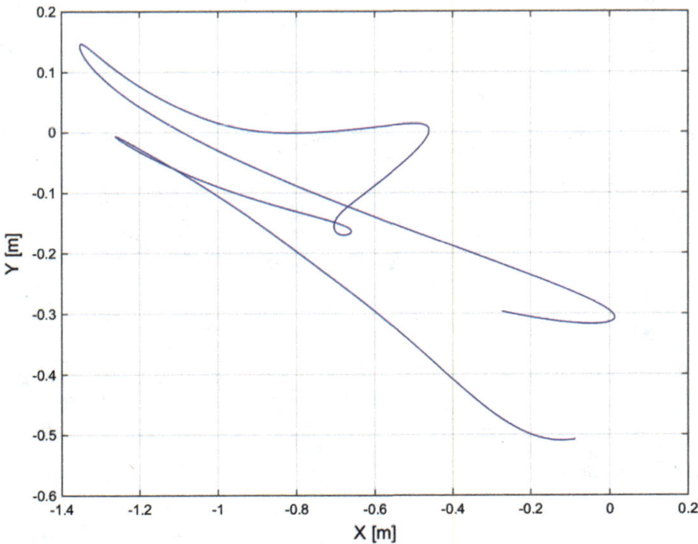

Fig. 3 Base motion during the end effector trajectory tracking

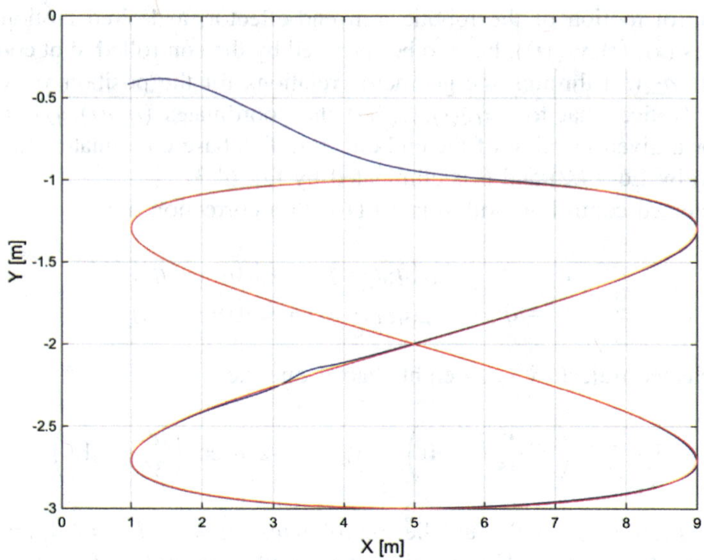

Fig. 4 Desired trajectory tracking by the end-effector

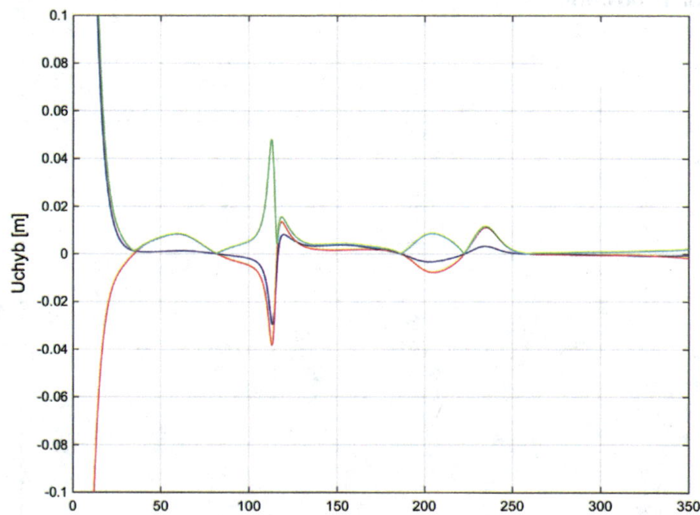

Fig. 5 End-effector tracking errors

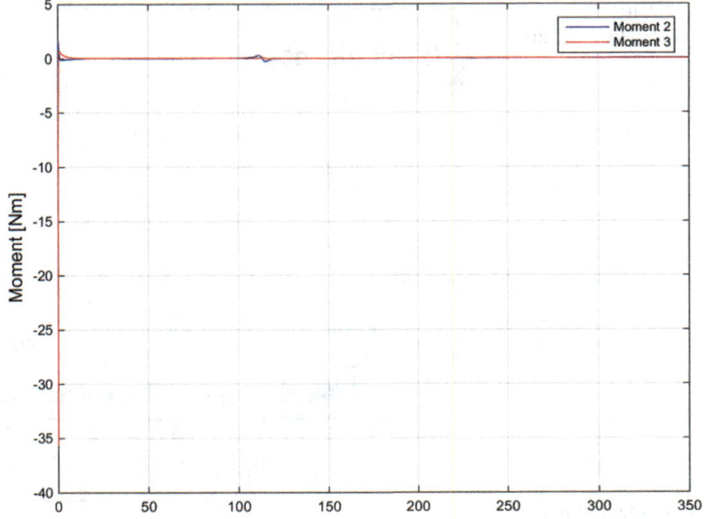

Fig. 6 Control torques in the arm joints

4.2 *Tracking a Desired Acceleration Change by a 2-D Spacecraft-Robotic Arm System End-Effector*

This simulation study demonstrates real capabilities of the GPME (2.4). The desired robotic arm end-effector performance specifies the acceleration. The programmed constraint is then

$$\sqrt{(\ddot{x}_E)^2 + (\ddot{y}_E)^2} - C(t) = 0 = \varphi, \tag{4.13a}$$

which differentiated and transformed to the joint space yields

$$\dddot{q}_1 a_1 + \dddot{q}_2 a_1 + A - \varphi\dot{\varphi} + \alpha\varphi = 0, \tag{4.13b}$$

with a_1, a_2 and A which are not functions of third order derivatives of q's. The acceleration is required to be constant, i.e. $C(t) = 0.55$. The term $\alpha\varphi$ is for the numerical solution of the constrained dynamics stabilization. The constraints are second order nonlinear nonholonomic. The differentiated form (4.13b), i.e. linear third order is easier to handle in the derivation of (2.4) and in simulations.

Notice, that for the program (4.13a) the Lagrange equations cannot be used.

Computed torque controller (4.9) with a PD is used for this simulation. Simulation results, i.e. the desired acceleration change of the end-effector and accelerations generated by this motion in the links are presented in Figs. 7 and 8.

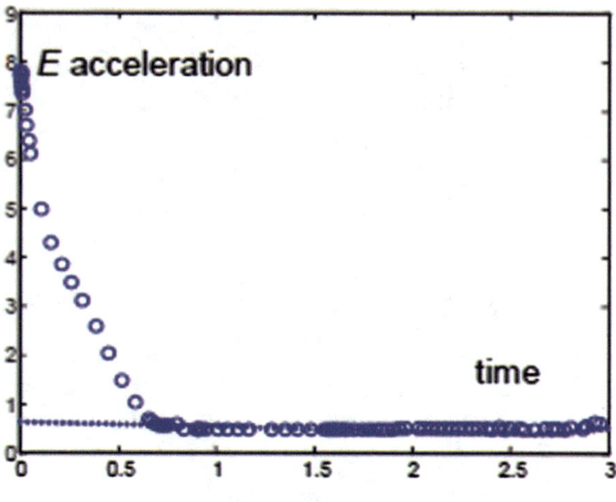

Fig. 7 End effector controlled acceleration versus time

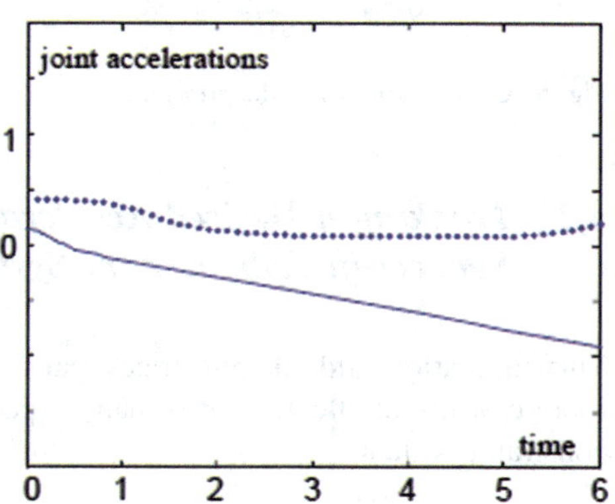

Fig. 8 Joint accelerations of the robotic arms versus time

5 Conclusions

The paper presents a model-based controller design for a robotic arm mounted on a spacecraft, which is to perform a desired motion. This desired performance is specified by constraint equations and the constrained dynamics and control dynamics are developed based upon one multi-purpose analytical dynamics modeling framework, which constitutes a basis for a development of an advanced control platform. Simulation studies for desired motion tracking, which are a trajectory and velocity changes for the manipulator end-effector, illustrate the theoretical development in modeling and the control platform applications. The control platform, originally developed for fully actuated system control models, provides a unified tool that enables planning and controlling a pre-specified unactuated system performance.

References

Castronuovo M (2011) Active space debris removal—a preliminary mission analysis and design. Acta Astronaut 105(9–10):848–859

Fantoni I, Lozano R (2002) Non-linear control for underactuated mechanical systems. Springer, London

Hervas JR, Reyhanoglu M (2013) Control and stabilization of a third-order nonholonomic system. In: Proceedings of the 13th international conference on control, automation and systems, ICCAS, pp 17–22

Jarzębowska E (2012) Model-based tracking control of nonlinear systems. Series: modern mechanics and mathematics. Taylor & Francis Group, Boca Raton

Jarzębowska E, Pietrak K (2014) Constrained mechanical systems modeling and control: a free-floating space manipulator case as a multi-constrained system. Rob Auton Syst 62:1353–1360

Jarzebowska E, Pilarczyk B (2015) Model-based control design for a free-floating space manipulator capturing debris. In: Proceedings of the 6th European conference for astronautics and space science, Cracow, Poland, 29 June–3 July

Jarzębowska E, Augustynek K, Urbaś A (2017) Computational derivation based reference dynamics model of a multibody system model with first order nonholonomic constraints. In: Proceedings ASME 2017 international design engineering technical conferences & computers and information in engineering conference, IDETC2017, Cleveland, Ohio, USA, 6–9 Aug 2017

Liang B, Xu Y, Bergerman M (1996) Mapping a space manipulator to a dynamically equivalent manipulator. Technical report CMU-RI-TR-96-33, Carnegie Mellon University

Liang B, Xu Y, Bergerman M (1998) Mapping a space manipulator to a dynamically equivalent manipulator. J Dyn Syst Measur Contr 120(1):1–7

Papadopoulos E (1990) The dynamics and control of space manipulator systems. Dissertation, Massachusetts Institute of Technology

Papadopoulos E, Dubowsky S (1991) Coordinated manipulator/spacecraft motion control for space robotic systems. In: Proceedings 1991 IEEE international conference on robotics and automation

Papadopoulos E, Dubowsky S (1991b) On the nature of control algorithms for free-floating space manipulators. Rob Autom IEEE Trans 7(6):750–758

Udwadia FE, Wanichanon T (2014) Control of uncertain nonlinear multibody mechanical systems. J Appl Mech 81(4)

Vafa Z, Dubowsky S (1990) The kinematics and dynamics of space manipulators: the virtual manipulator approach. Int J Robot Res 9(4):3–21

References

[illegible]

Detection and Decoding of AIS Navigation Messages by a Low Earth Orbit Satellite

Roman Wawrzaszek, Marcin Waraksa, Maciej Kalarus, Grzegorz Juchnikowski and Tomasz Górski

1 Introduction

1.1 Automatic Identification System

The Automatic Identification System (AIS) has been recognized by international authorities as an effective tool for shipping safety and security management on a world-wide scale. With the growing role of the AIS system in the global maritime picture awareness (Moua et al. 2010) and new requirements imposed by maritime authorities, coastguards and naval authorities from countries with their own chains of the AIS base stations (PSS—Physical Shore Station) and relevant telecommunication infrastructure, the AIS is under constant development. In the past, new services have been implemented, such as the introduction of AIS class B transponders (dedicated for non-SOLAS[1] units like small fishing vessels, pleasure crafts, etc.), aids-to-navigation (AtoN) monitoring, synthetic/virtual AtoN's,

[1]SOLAS—International Convention for the Safety of Life at Sea.

R. Wawrzaszek (✉) · M. Kalarus · G. Juchnikowski
Space Research Centre PAS, Bartycka 18A, 00-716 Warsaw, Poland
e-mail: wawrzasz@cbk.waw.pl

M. Kalarus
e-mail: kalma@cbk.waw.pl

G. Juchnikowski
e-mail: grzesiek@cbk.waw.pl

M. Waraksa
Gdynia Maritime University, Morska 81-87, 81-225 Gdynia, Poland
e-mail: m.waraksa@we.am.gdynia.pl

T. Górski
Creotech Instruments S.A., Gen. L. Okulickiego 7/9, 05-500 Piaseczno, Poland
e-mail: tomasz.gorski@creotech.pl

J. Sasiadek, *Aerospace Robotics III*, GeoPlanet: Earth and Planetary Sciences,
https://doi.org/10.1007/978-3-319-94517-0_4

Emergency Position Indicating Radio Beacons (EPIRB) equipped with AIS transponders or AIS-SART (Search and Rescue Transponders). The last two services (EPIRB-AIS and AIS-SART) have been introduced for quick and precise life-rafts or man-overboard localization at sea.

Another significant factor affecting AIS development is the rapidly growing number of Application Specific Messages (ASM) exchanged between dedicated end-users. ASMs are not directly related to the core functionality of the AIS.

The introduction of the abovementioned services and devices caused a significant VHF Data Link (VDL) load—in several hotspots, the VDL load is higher than 64% (northern part of the Gulf of Mexico) and 40% in the territorial waters of Japan and South Korea (CEPT ECC 2013).

Access to the transmission medium in Automatic Identification System is based on one of four available access schemes: SO-TDMA (Self Organized Time Division Multiple Access), IT-DMA (Incremental Time Division Multiple Access), RA-TDMA (Random Access Time Division Multiple Access) and FA-TDMA (Fixed Access Time Division Multiple Access) (ITU 2014). The essential role of the VHF Data Link (VDL) is to provide a communication link for data exchange in the marine VHF band (VHF channels 87B, 88B—161.975 and 162.025 [MHz] respectively—(ITU 2014))—dedicated to the Automatic Identification System. In a single physical channel 2250 (numbered form 0 to 2249) time slots has been defined. A single time slot lasts 26.66 ms (ITU 2014), so the total AIS frame lasts 60 s. Defined in the (ITU 2014) set of AIS messages can occupy from 1 up to 5 consecutive time slots, where single slot messages are the dominating type.

1.2 AIS—Satellite Segment

Access schemes, described in Sect. 1.1, has been designed for the terrestrial AIS segment only. Those access schemes work fine with ship-to-ship or ship-to-shore communication. However, in case of the satellite segment of the AIS, the efficiency of implemented access schemes is heavily degraded due to the high ratio of the AIS packets collisions. This is caused mainly by a high number of active AIS transponders being in the satellite field-of-view (FOV) (Cervera et al. 2011; Eriksen et al. 2006). Transponders from different terrestrial AIS service areas, being in the satellite FOV, are not synchronized between themselves or with the satellite.

1.2.1 AIS Packets Collision

The most important source of AIS packet collisions during space detection is an access scheme implemented into the AIS system. This problem does not appear in the terrestrial AIS segment, where synchronization between transponders being in the particular terrestrial AIS service area is implemented (typically within 40 nm range)—Fig. 1.

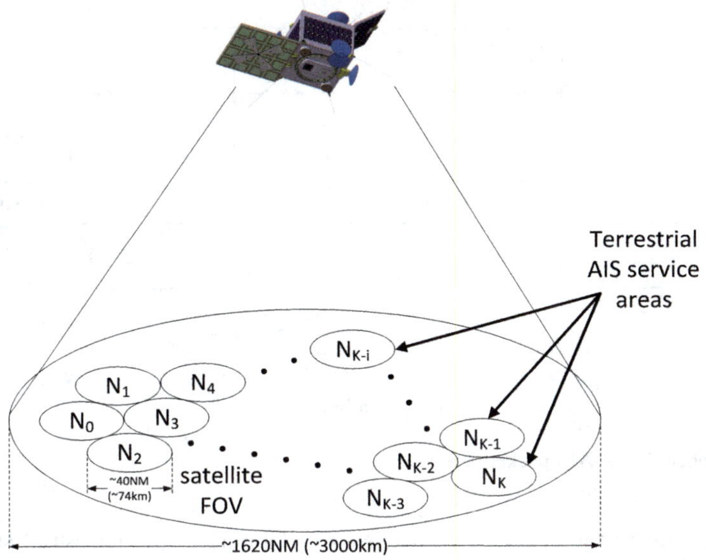

Fig. 1 AIS service areas in the satellite field-of-view

From perspective of the AIS receiver, installed at a satellite flying on the Low Earth Orbit (LEO), it is possible to have 15,000 active transponders within the satellite FOV in the zones of a high density of marine traffic. The mathematical model for the analysis of the probability of the AIS packets detection was proposed by the research team working under the leadership of G. K. Høye at FFI—the Norwegian Defence Research Establishment (Høye et al. 2008). The probability of the AIS packets detection with respect to vessel numbers and satellite FOV size is presented in Fig. 2. In the case of low density marine traffic zones—with less than 1000 active AIS transponders in the satellite FOV, packet collisions do not have a significant influence on detection probability. This problem appears in high density ship traffic zones and strongly reduces the probability of AIS packet detection.

$$P = 1 - \left[1 - \exp\left(-\frac{(1+s)\cdot N_{tot}}{37,5\cdot n_{ch}\cdot \Delta T}\right)\right]^{\frac{T_{obs}}{\Delta T}} \tag{1}$$

An Eq. (1) defining the AIS packet detection probability P, where:

s overlap factor (depending on satellite altitude and FOV)
N_{tot} vessel number in the satellite FOV
n_{ch} number of AIS channels ($n_{ch} = 2$)
ΔT average reporting period ($\Delta T = 6$ s)
T_{obs} observation time

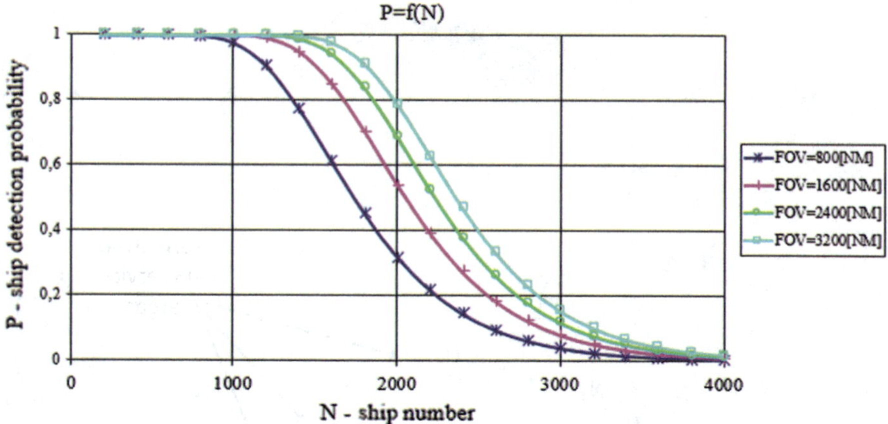

Fig. 2 Probability of AIS packet detection

To minimize problems of AIS packets overlapping in the terrestrial segment a 24-bit length buffer has been defined, from witch a 14-bit length propagation time delay difference segment has been separated. This segment protects AIS packets from partial or full overlapping for transponders being within 236 nm range, sufficient in the terrestrial AIS segment.

Another factor causing AIS packets collision is the Faraday effect, causing the rotation of the signal polarization plane. The angle of the plane rotation depends on signal frequency, elevation angle, magnetic flux density and electron distribution density in the ionosphere.

An additional factor causing AIS packets overlapping and collisions is the Doppler effect. Vessel speed and course can be neglected with respect to satellite speed. Doppler frequencies for AIS signals vary form 0 Hz (NADIR) to 4.1 kHz (elevation 0° and satellite orbit 500 km), as shown in Fig. 3.

Atmosphere attenuation (less than 0.05 dB) of the AIS signal, as well as multipath propagation appearing at low elevation, are negligibly small and do not have practical influence on AIS packets' detection probability.

1.2.2 Improvements of the Probability of AIS Packets Detection

The basic solution for collision reduction has been introduced during the World Radio Conference 2012 (WRC2012)—two additional frequency channels (75 and 76 of the marine VHF band–156.775 and 156.825 [MHz] respectively) for satellite AIS segment have been introduced. Those two additional channels are used for Message 27 broadcast only and neither ship AIS stations nor PSSs are capable of receiving them.

Apart from the two additional channels (75 and 76) assigned for the satellite AIS segment only, end-users require information delivered via AIS1 and AIS2 channels (87B and 88B, respectively). Research shows that approximately 80–85% of AIS

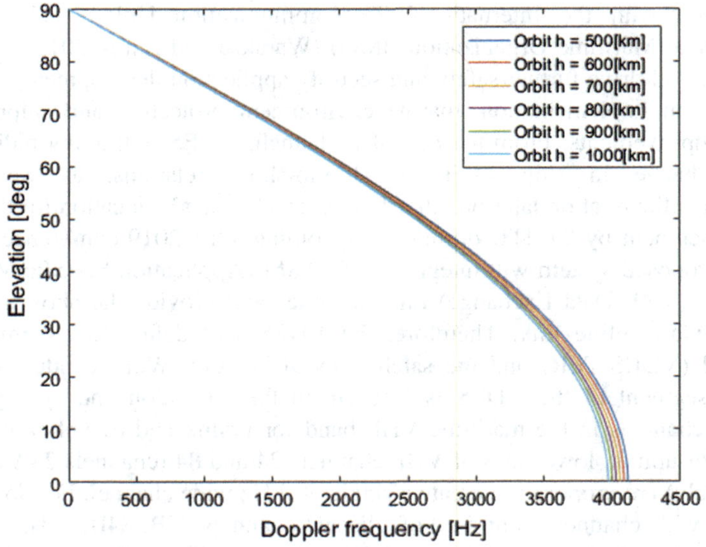

Fig. 3 Doppler frequency for f_0 = 162 MHz, satellite orbit H_{sat} = (500–1000) km

packets received by the satellite are lost due to packets collisions (ITU 2009) at the satellite receiver. To fulfill SAT-AIS end-users requirements and to deliver to them as much AIS packets as possible, it is necessary to implement advanced signal processing algorithms on the raw signal in the form of in-phase and quadrature signal components received by the satellite. Commonly used algorithms for the AIS decollisioning are: Maximum Likelihood Sequence Estimation, Forward Linear Prediction (FLP), Backward Linear Prediction (BLP) or Forward/Backward Linear Prediction (FBLP) as well as Blind Source Separation. The mathematical description of these methods is beyond the scope of this paper. Important disadvantages of the abovementioned methods of signal processing are their computational complexity and the time needed for signal analysis. In most cases, the time needed for signal analysis and processing is longer than the information remains useful for the end-user. The only exception in the use of post-processed SAT-AIS information is the statistical analysis of marine traffic.

1.3 VHF Data Exchange System

Due to the increasing VDL load of the AIS system having a destructive influence on the system itself and its stability, a new solution for the maritime community is to be put in place. The International Association of Marine Aids to Navigation and Lighthouse Authorities (IALA) is the driving force behind the new solution, the VHF Data Exchange System (VDES). System development is performed in close cooperation with the International Telecommunication Union (ITU) and the International Maritime Organization (IMO) (Waraksa and Zurek 2015).

VDES will drive further safety and security application development, as well as e-Navigation implementation, marine environment protection and shipping efficiency improvements. From the day of its launch, VDES will allow bidirectional data exchange in ship-to-ship or ship-to-shore relations, as well as in ship-to-satellite relation later on, after frequency channels' allocation for the VDES satellite segment by the ITU during the upcoming WRC2019 conference.

The proposed system will integrate AIS, ASM (Application Specific Messages) and VDE (VHF Data Exchange) functionalities and provide the abovementioned services via satellite link. Therefore, the VDES will define two segments: the terrestrial (VDES-TER) and the satellite (VDES-SAT). Work conducted on the satellite segment of the VDES is focused on the allocation and aggregation of adjacent channels in the maritime VHF band for uplink and downlink communication. For uplink, lower-legs of VHF channels 24 and 84 (channels 24A and 84A, respectively) are proposed to create a single 50 kHz wide channel. For downlink, a 150 kHz wide channel, comprised of adjacent channels 24B, 84B, 25B, 85B, 26B, 86B, is proposed. Data transfer in the VDES-SAT segment will be provided in two manners:

- broadcast transmission—for all vessels being in the satellite FOV
- multicast transmission—for single flag or ship owner fleet or vessels sailing through the particular area of interest

Such an organization of the transmission allows for the effective utilization of the limited physical resources allocated to the maritime VHF band. In the VDES-SAT maximum downlink throughput is 307.2 kbps.

What is more, the satellite segment of the VDES will provide new digital services to the rapidly growing non-SOLAS fleet without the need of the allocation of significant financial resources into high-end satellite communication systems. Newly designed services require an uplink channel for quality-of-service (QoS) measurements such as data packets repetition requests, information requests and the confirmation of a successful reception (Waraksa and Zurek 2015) (Table 1).

The VDES is taken into consideration as the SAT-AIS-PL main payload. Depending on the decision to be taken during the World Radio Conference 2019, a final decision regarding the VDES transceiver will be made by the members of project consortium. In the case of the VDES-SAT frequency allocation during the WRC2019, a full transceiver will be implemented. In the opposite case, the VDES-TER receiver will only be implemented on the SAT-AIS-PL satellite. It is important to underline that full VDES functionality will be implemented as soon as physical resources will be assigned to the VDES-SAT.

Table 1 Marine VHF band channels assigned for VDES to provide AIS, ASM, VDE and SAT services (ITU 2015)

Channel no. (radio regulation, appendix 18)	Ship and shore station frequency [MHz]	
	Ship station (ship-to-shore) SAT-AIS Ship station (ship-to-satellite)	Shore station Ship station (ship-to-ship) Satellite-to-ship
AIS 1	161.975	161.975
AIS 2	162.025	162.025
75 (SAT-AIS)	156.775 (Ship station, Tx only)	n/a
76 (SAT-AIS)	156.825 (Ship station. Tx only)	n/a
27B (ASM 1)	161.950 (27B)	161.950 (27B)
28B (ASM 2)	162.000 (28B)	162.000 (28B)
24/84/25/85 (VDE 1)	25/100 kHz	25/100 kHz
	(24/84/25/85, adjacent lower-leg of duplex channels)	(24/84/25/85, adjacent upper-leg of duplex channels)
	Ship-to-shore	Ship-to-ship, shore-to-ship
	Ship-to-satellite	Satellite-to-ship under restricted conditions
24	157.200 (24A)	161.800 (24B)
84	157.225 (84A)	161.825 (84B)
25	157.250 (25A)	161.850 (25B)
85	157.275 (85A)	161.875 (85B)
26/86 (SAT 1/VDE 2)	25/50 kHz	25/50 kHz
	(26/86, adjacent lower-leg of duplex channels)	(26/86, adjacent upper-leg of duplex channel)
	Ship-to-satellite/shore	Satellite/shore-to-ship
26	157.300 (26A)	161.900 (26B)
86	157.325 (86A)	161.925 (86B)

Notation xxA—lower leg of the VHF xx channel; xxB—upper leg of the VHF xx channel

2 AIS Messages Detection and Decoding

This chapter describes the algorithm for the detection and decoding of AIS messages implemented and tested by its authors. To understand the demodulation and decoding processes, one should first understand how the AIS message is coded and modulated. This process is well described in (Dembovskis 2015). A short summary is provided below.

The AIS message is a series of bits containing: 8 bit ramp up, 24 bit long training sequence (010101…0101), the HDLC marker (01111110)—start flag, 168 bit long AIS data segment, 16-bits of CRC, another 8 bit long HDLC marker (end flag) and 24 bit long buffer (stuffing distance delays, repeater delay and jitter). Bit stuffing is performed on the part of the message between two HDLC markers. The resulting stream of bits is NRZI (Non-Return-to-Zero-Inverted) coded, which means that output changes state when '0' is on input, and stay unchanged when '1' is on input. The bits are then transformed into a continuous signal $\varphi_O(t)$, which, during the bit interval (1/9600 s) linearly increase or decrease by π/2, depending on the value of the bit. This signal is then filtered with a Gaussian low pass filter giving $\varphi(t)$. Then the two signal components (in-phase—*I* and quadrature—*Q*) are produced (2) and (3):

$$I(t) = \cos(\varphi(t)) \tag{2}$$

$$Q(t) = \sin(\varphi(t)) \tag{3}$$

The composed signal, *I*, *Q* is modulated with a carrier frequency f_c (161.975 MHz or 162.025 MHz), amplified and transmitted. This modulation can be described with a simplified Eq. (4).

$$TX(t) = I(t)\cos(2\pi f_c t) - Q(t)\sin(2\pi f_c t) = \cos(2\pi f_c t + \varphi(t)) \tag{4}$$

This description of modulation may look differently than the standard explanation of GMSK modulation, which provided a foundation for AIS, but, in fact, these two descriptions are equivalent. To see this, please assign f_c in the Eq. (4) for *TX*(t) to $f_c = 1.25 f_b$, where $f_b = 9600$ is the bit rate. In this case, the phase of the cos(·) component during one bit interval changes by 2π or by 3π, depending on the bit value. This is in accordance with the GMSK modulation definition.

Further processing takes place in a receiver. Let us assume that the received signal is the same as the transmitted one:

$$RX(\mathrm{t}) = TX(\mathrm{t}) \tag{5}$$

Changes of the amplitude and a delay can be neglected. The quadrature receiver, e.g. Software Defined Radio (SDR), produces a composed signal I_i(t), Q_i(t), which corresponds to *I* and *Q* signal components mentioned before, but modulated with an intermediate frequency f_i. Depending on the SDR configuration, this intermediate frequency can be positive, negative, or even zero, as it will be shown in an example. The intermediate frequency is selected at a value low enough to convert the analog signal to a digital one. The signal takes the form of a series of numbers (I_k, Q_k), sampled with frequency f_s. It is convenient to treat *I* and *Q* as real and imaginary components of a complex signal Z_1 (6).

$$Z_{1,k} = I_k + jQ_k = A_k e^{j((2\pi f_i/f_s)k + \varphi_k)} \tag{6}$$

where

$$\varphi = \varphi_m + \varphi_e \tag{7}$$

The component φ_m is a phase deviation, which carries information about the AIS message. Ideally, the φ_m phase in the interval of one transmitted bit linearly increases or decreases by $\pi/2$, depending on the bit value. The real shape of φ_m is smoothed due to the Gaussian filtering on the transmitter side and due to the limited frequency band. The φ_e component is a sum of disturbances. One of the disturbances is the Doppler shift, which is seen as a φ_e linear increase or decrease in time. The transient amplitude, A_k carries no useful information, except in the case when $A_k = 0$ (when $I = Q = 0$). Then the phase φ_k cannot be determined.

The first step of the AIS packets decoding is a signal phase φ extraction from the input signal Z_1. This is done in a standard manner, used in many applications. The quadrature demodulation is described with the Eq. (8):

$$Z_{2,k} = Z_{1,k}e^{-j(2\pi f_i/f_s)k} = A_k e^{j\varphi_k} \tag{8}$$

In Eq. (8), an additional, unknown constant phase shift has been intentionally omitted. It can be neglected because a constant phase shift does not have an influence on the decoding process. All useful information has been demodulated, but the signal contains many unwanted components of higher frequencies which should be filtered out. Filtering is performed by a low-pass, 5th order Butterworth filter with 10 kHz cut-off frequency, giving a complex Z_3 signal as a result—now the phase φ of the Z_3 signal can be computed as follows (9):

$$\varphi = Arg(Z_3) \tag{9}$$

It is convenient to use the function atan2(Q, I) for Z_3 signal phase computation. The signal phase is an ambiguous value, e.g. $\varphi = 3\pi/2$ and $\varphi = -\pi/2$ signify the same signal phase. Therefore, particular care should be taken choosing a signal phase value for further computations—the chosen phase value should not differ by more than π with from the previously used phase value. This continuity is necessary because we will soon compute a time derivative of the signal phase. Before derivation, the signal is again filtered with a low-pass filter—this time a Gaussian filter with a bandwidth–time product of BT = 0.4 (ITU 2014). The second filtering necessity has been empirically proven. As it was already mentioned before, the phase cannot be determined when both I and Q signal components equal zero. The implementation described here assumes the last calculated phase to be valid.

Time derivative of this phase is calculated using the following Eq. (10):

$$S_k = (\varphi_k - \varphi_{k-1})T_{bit} \tag{10}$$

where

$$T_{bit} = f_s/9600 \tag{11}$$

is a bit interval expressed as a number of samples. Multiplying by T_{bit} scales the derivative S_k, so that the nominal AIS signal has the amplitude of $\pm\pi/2$.

The next step is the detection and removing a constant bias from the series S_k (e.g. Doppler shift). The S_{avg} value is calculated as a sum of last d samples of S_k, divided by d. The delay d should cover a multiplicity of four bit time intervals. We use the following value for d (12) here:

$$d = 12T_{bit} \tag{12}$$

Keeping the multiplication by four bit time intervals is important because during the training sequence, the starting AIS message signal S_k has a period of four bit time intervals. The average value is then subtracted from S_k, giving S_k^0, which is unbiased. An update of the average stops and the S_{avg} is put on hold as soon as the start of the message is detected. It must be done that way because the signal of the message is not as symmetrical as during the training sequence, and continuing the calculation of the average could deteriorate it.

An implementation note: to calculate the sum of S_k from the current one back to S_{k-d+1}, it suffices to perform simple iteration (13):

$$sum_k = sum_{k-1} + S_k - S_{k-d} \tag{13}$$

The delayed S_{k-d} is taken from a cyclic First In, First Out (FIFO) queue of size d. The FIFO is used in a (14) way:

$$\begin{aligned} S_{k-d} &= fifo[ptr] \\ fifo[ptr] &= S_k \\ ptr &= (ptr + 1) \bmod d \end{aligned} \tag{14}$$

The time moments of bit beginnings are determined during the bit synchronization procedure. For this purpose, the crossing of zero by S_k^0 is used. Before the message begins, each zero crossing is detected and the potential bit beginning defined exactly the moment of zero crossing. The bit time timer is then reset and starts counting modulo T_{bit}. This is performed differently inside the message body: every zero crossing synchronizes the bit time timer, even if not so strictly, i.e. the timer is updated only by a fraction of the difference between its current value and 0 or T_{bit} (whichever is closer).

Single sample of S_k^0, taken in the middle of the bit interval period, does not contain any information. Only the relation between the current and previous sample carries information. When both samples are of the same sign, positive or negative, then the bit message is a logical '1'. Similarly, when the samples have different signs, then the bit message is a logical '0'. This procedure reverses the NRZI mechanism applied before transmitting the message.

Detection of the beginning of the AIS message is performed based on bit stream processing. Bits are shifted in a shift register from the least significant bit (LSB) side. The content of the register is processed by a Finite State Machine (FSM). The FSM has the following steps to accomplish:

1. Training sequence "…01010101010" detection—in this moment, the S_{avg} should be calculated and held.

2. HDLC marker "…0101111110" detection. The following bit will be the first bit of the AIS message.
3. Collecting AIS message bits in a buffer. The overflow of the buffer resets the FSM.
4. Detecting the terminating HDLC marker, which resets the FSM and triggers further steps of message decoding.

Every deviation of the shift register content from the expected value leads to an FSM reset.

Further steps are known. First, one must reverse bit stuffing and check that the message contains a number of bits that is a multiplication of 8. Next, one should reject the HDLC markers, verify the CRC, and then reverse the bits' order in every 8-bits' length word. The final step should extract the message type code and, depending on the type code, perform the extraction of the remaining data. These steps are well described e.g. in (Dembovskis 2015), (ITU 2014).

The algorithm works in the streaming mode. It does not need memory buffering for the input of data for the whole message. Therefore, it can easily be implemented into a FPGA, but for the purposes of the presented research, a dedicated C++ tool has been developed and tested.

3 Experimental Data Analysis

3.1 Terrestrial Data Set Analysis

An example of terrestrial AIS data set for further analysis has been registered at the Gulf of Gdańsk, close to the Gdynia Harbor. The AIS signal has been registered using a Software Defined Radio (SDR) unit, in that case the RTL-SDR v3 Bath 2 module. The signal duration was 30 s, the sampling frequency was 250 kHz and the intermediate frequency was 50 kHz for AIS channel A, and 0 kHz for AIS channel B. Using a simple filtering of the signal by a moving average, several AIS packets were easily identified (Fig. 4).

The 4th message in the sample data set, characterized by the highest signal amplitude, is an AIS channel B message (Fig. 4). The in-phase and Quadrature signal components of the waveform, which correspond to this particular message,

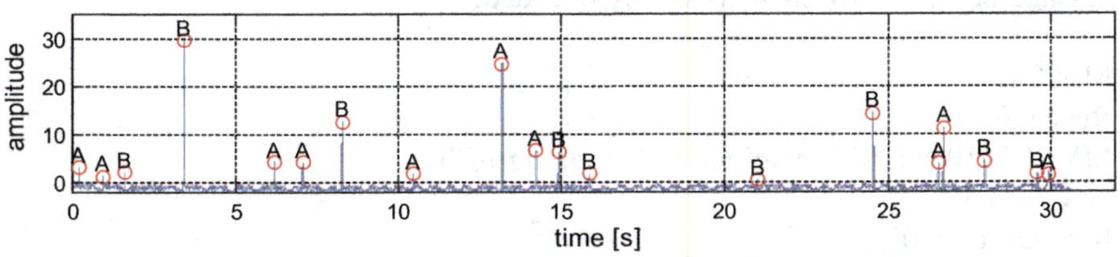

Fig. 4 AIS messages detected by the means of a moving average

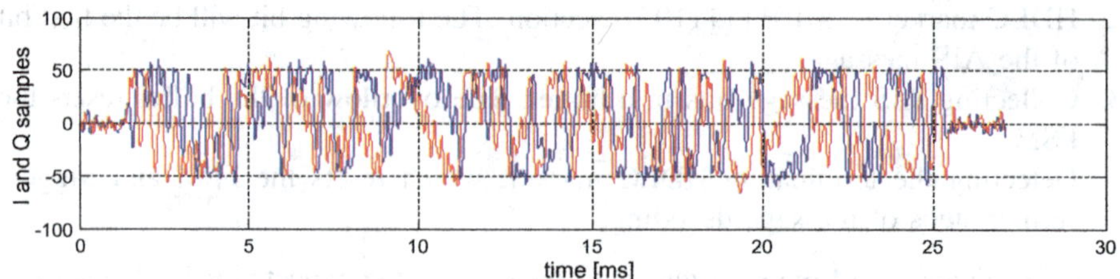

Fig. 5 I and Q signal components of AIS message registered at Gulf of Gdańsk

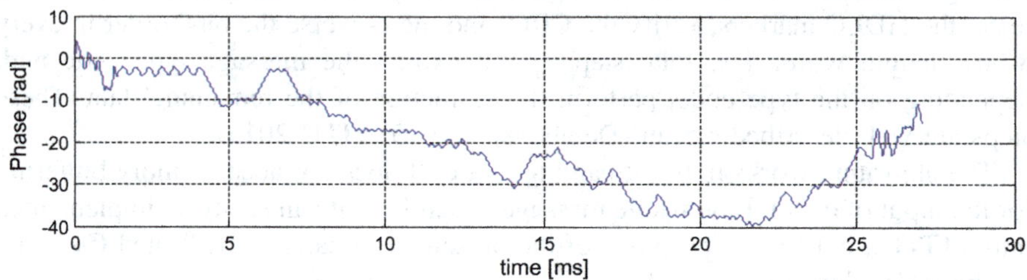

Fig. 6 The determined signal phase

are presented in Fig. 5. The further analysis of this signal resulted in a determination of the phase variation (Fig. 6).

By the implementation of the methodology described in Sect. 2, the following bit string has been identified in this message upon its NRZI decoding:

0111010101010101010101010101**01111110**0010000001111100000011110010000
1100000011000000000000000000001100000101010110000111001001011111 00
0000101100100111010011011111001101110111000001101100000000000001000
00100111101100011100101000**0111111 0**11

In the presented string, the beginning and the end of the AIS message frame have been marked in bold. Directly before the commencement marker, the "training" sequence, made of twelve consecutive pairs of "01", is clearly visible (Fig. 7).

After destuffing and bit swapping following the AIS standard (ITU 2014), the message below has been successfully decoded:

'Msg Type:1'
'Repeat:0'
'MMSI:261036120' (*vessel name "Oceanograf"*)
'Nav State:0'
'Rate Of Turn:0'
'Speed:0.1'

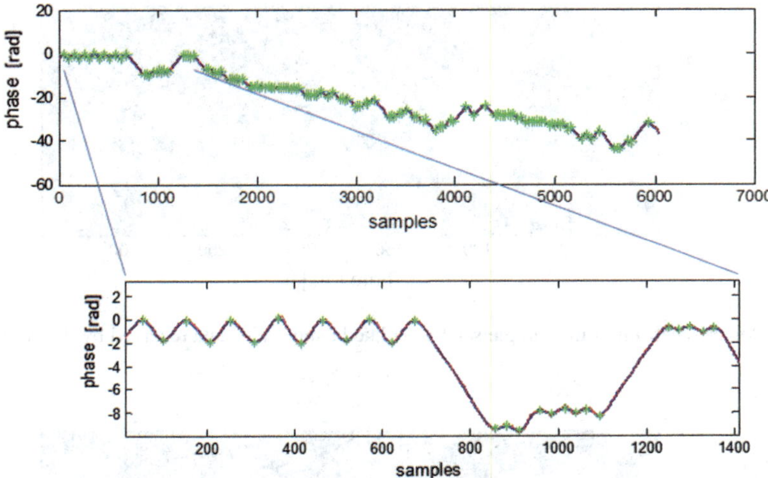

Fig. 7 The phase of the AIS message after smoothing and local peak identification (green markers). The bottom plot shows a close-up of the training sequence and the message beginning mark

'Position Accuracy:1'
'Longitude:18.5167'
'Latitude:54.5364'
'Course Over Ground:329.1'
'Heading:58'
'Time Stamp:4'
'RAIM:0'
'Communication State:82046'

3.2 Low Earth Orbit Data Set Analysis

The data sets collected on-orbit (Fig. 8) and used in testing were registered by the AAUSAT3 satellite built at Aalborg University (Jessen et al. 2009). The data has been shared through (AAUSAT3 2017). AAUSAT3 is a CubeSat, Low Earth Orbit satellite launched on the 25th of February 2013 at 13:31CET. Its main payload is an SDR AIS receiver. AAUSAT3 is on the polar orbit, at the altitude of 781 km. The parameters of two time series of obtained data are the following: duration is 346 ms, sampling frequency is 758,272 Hz and intermediate frequency is 175 kHz for channel B and −225 kHz for channel A.

Figures 8 and 9 show the power spectrum of two obtained data sets. An analysis of these results shows that there are signals in set 2 which correspond to AIS channel B and probably contain AIS messages. Time moments indicated by orange

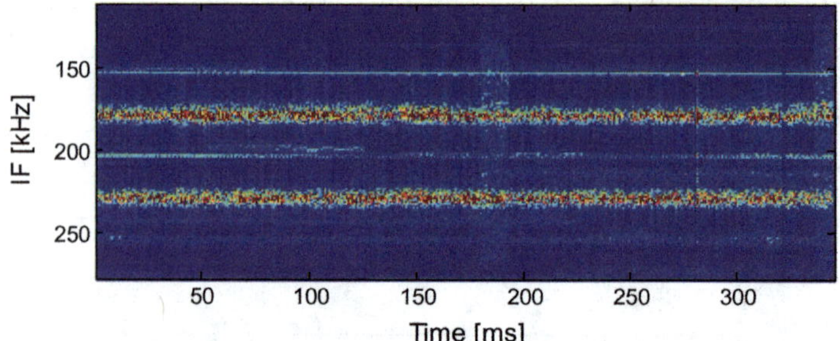

Fig. 8 Power spectrum of the sample set 1 calculated using FFT, data received by the AAUSAT3 satellite

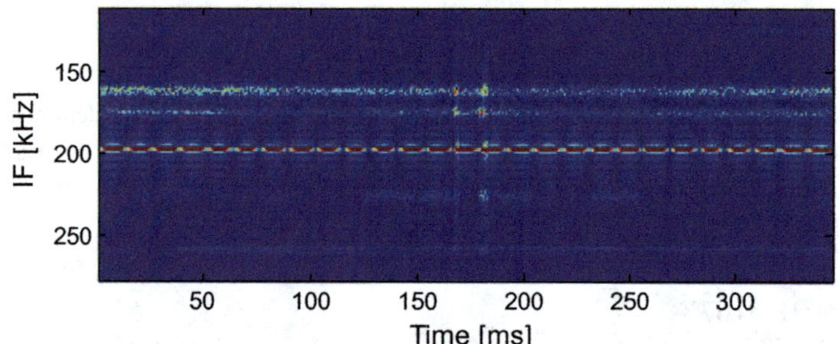

Fig. 9 Power spectrum of the sample set 2 calculated using FFT, data received by the AAUSAT3 satellite

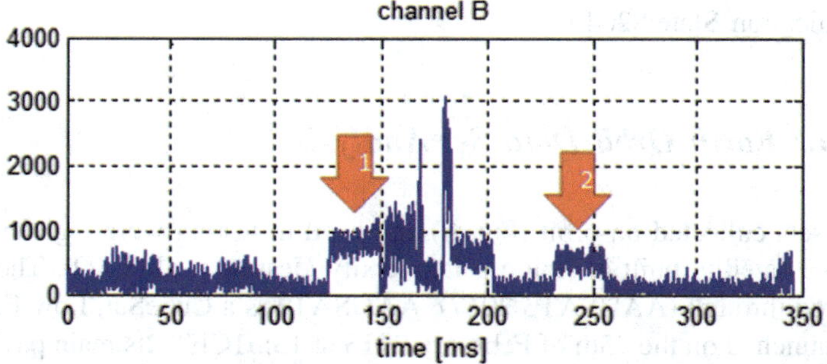

Fig. 10 An AIS signal extracted from data set 2. Frequency corresponds to the AIS2 channel. Location of the analyzed AIS messages marked by arrows. Data received by the AAUSAT3 satellite

arrows on Fig. 10 were chosen for further analysis. These signals' further analysis has been performed in accordance with methods described in Sect. 2 and similar to the methods used in the case of the terrestrial AIS data set (Fig. 11).

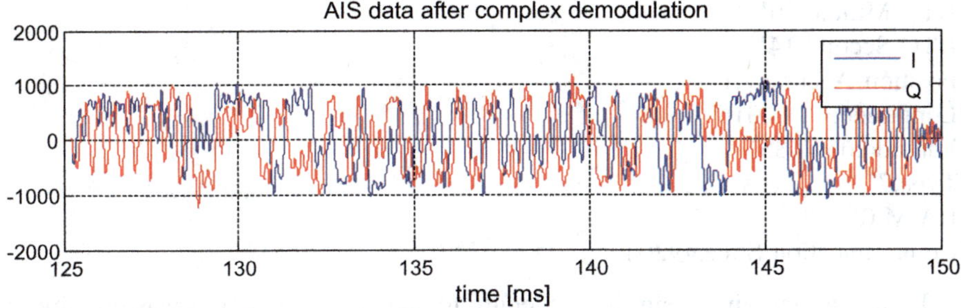

Fig. 11 I and Q components of AIS signal registered on the orbit. Signal corresponds to the message marked by arrow 1 in Fig. 10

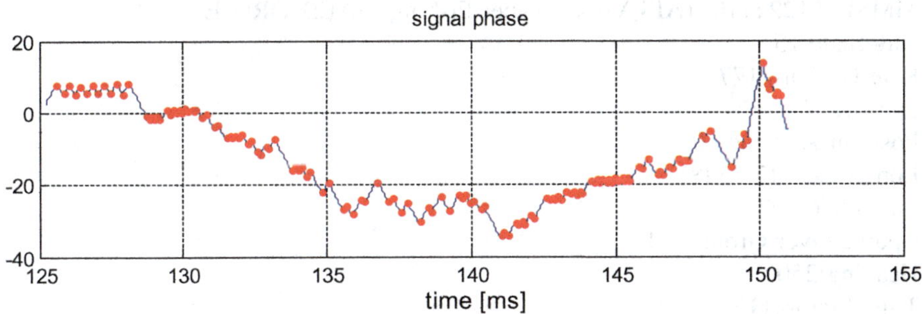

Fig. 12 The phase of the AIS message signal after smoothing and local peak identification (red markers). The signal corresponds to the message marked by arrow 1 in Fig. 10

The bit stream determined based on the phase presented in Fig. 12 is as follows:

10101010101010101010101**01111110**00001000000000001001100110000001001
101001011111000010011101011110010110011101110011010111011001101101
100010010011111000110000100111000000010000001100000000000000110010 1
1000100100001111001**01111110**111001111011111

In the presented string, the beginning and end frame markers are marked in bold. The decoded message was as follows:

'Msg Type:4' AIS BASE STATION REPORT
'Repeat:0'
'User ID:2515019' (Base Station on the Island)
'UTC Year:2014'
'UTC Month:4'
'UTC Day:15'
'UTC Hour:10'

'UTC Minute:26'
'UTC Second:14'
'Position Accuracy:1'
'Longitude:−18.0016'
'Latitude:65.7485'
'Type:7'
'RAIM:0'
'Communication State:49204'

Thanks to applying a similar procedure to signal 2 (Fig. 12), it was possible to successfully decode the following message:

'Msg Type:3' Position report
'Repeat:1'
'MMSI:251224110' (AIS Vessel Type: Fishing; SAEBJORG EA-184)
'Nav State:15'
'Rate Of Turn:177'
'Speed:1.3'
'Position Accuracy:0'
'Longitude:−17.9348'
'Latitude:67.9722'
'Course Over Ground:74.4'
'Heading:250'
'Time Stamp:11'
'RAIM:0'
'Communication State:82768'

4 Conclusions

In this paper, we present our attempt at detection and decoding of AIS messages. The main task was to detect and decode messages collected by the AAUSAT3 satellite at the Low Earth Orbit. This analysis is based on the data collected by a terrestrial receiver. As the data collected in nominal conditions, the signals' parameters were not affected by significant interferences and thus comply with the AIS standard. This part of the work served us to verify our understanding of the AIS signal structure, as well as the demodulation and decoding processes. In the second part of the work, we have analyzed data collected by the AAUSAT3 satellite. When we used a method similar to the terrestrial case example, we correctly detected two AIS messages (of types 3 and 4). Nevertheless, our analysis confirmed many issues and difficulties of on-orbit collected data processing. As an example, an initial analysis of data set 1 can lead to the conclusion that unidentified signals in channel A and B are the results of interferences formed by overlapping AIS packets and possibly disturbed by other signals. Looking at power spectral density (Fig. 4), one can also notice the presence of several strong signals of an unknown origin, of

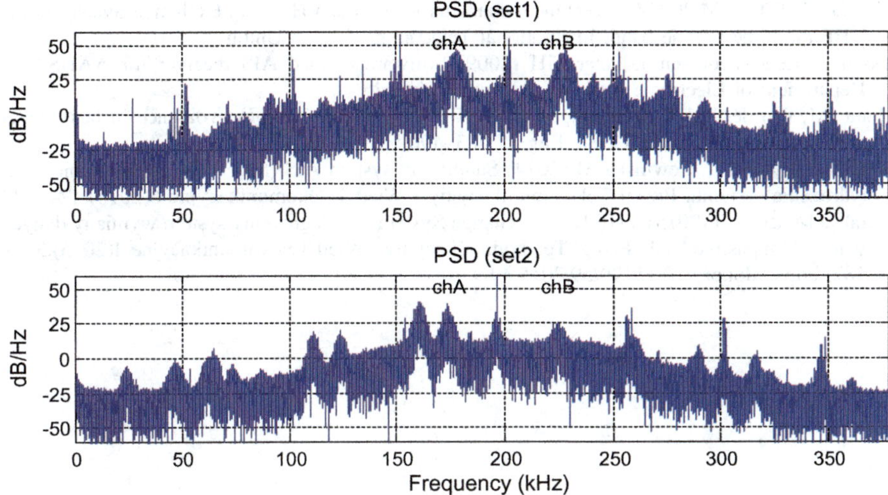

Fig. 13 Power spectral density of two data sets registered by AAUSAT3

frequencies close to those of the AIS channels. This may be a potential source of problems for the SAT-AIS message detection (Fig. 13).

Acknowledgements SAT-AIS-PL project is funded by the European Space Agency under awarded grant 4000116953/16/NL/Cbi—"SAT-AIS-PL Phase A".

Special thanks to the AAUSAT3 team and Aalborg University for sharing raw AIS I/Q data samples, which have been used for this analysis.

References

AAUSAT3 (2017) http://www.space.aau.dk/aausat3/index.php?n=Main.Getdata. Accessed on 6 June 2017

CEPT ECC (2013) CPG PTC(13) INFO 16—Information paper on VHF Data Exchange System (VDES). Electronic Communications Committee

Cervera MA, Ginesi A, Eckstein K (2011) Satellite-based vessel Automatic Identification System: a feasibility and performance analysis. Int J Satell Commun Network 29(2):117–142

Dembovskis A (2015) AIS message extraction from overlapped AIS signals for SAT-AIS applications. Dissertation, University of Bremen

Eriksen T, Høye G, Narheim B, Meland BJ (2006) Maritime traffic monitoring using a space-based AIS receiver. Acta Astronaut 58:537–549. https://doi.org/10.1016/j.actaastro.2005.12.016

Høye G, Eriksen T, Meland BJ, Narheim B (2008) Space-based AIS for global maritime traffic monitoring. Acta Astronaut 62(2–3):240–245. https://doi.org/10.1016/j.actaastro.2007.07.001

ITU (2009) ITU-R M.2169—Improved satellite detection of AIS. International Telecommunication Union

ITU (2014) ITU-R M.1371-5—Technical characteristics for an Automatic Identification System using Time Division Multiple Access in the VHF maritime mobile frequency band. International Telecommunication Union

ITU (2015) ITU-R M.2092-0—Technical characteristics for a VHF Data Exchange System in the VHF maritime mobile band. International Telecommunication Union

Jessen T, Ledet-Pedersen J, Peter MH (2009) Software defined AIS receiver for AAUSAT3. Department of Electronic Systems, Aalborg University

Moua JM, van der Tak C, Ligteringen H (2010) Study of on collision avoidance in busy water ways by using AIS data. Ocean Eng 37:483–490

Waraksa M, Żurek J, Lewińska M (2014) Satelitarny system AIS (SAT-AIS) rozwój i pierwsze wdrożenia systemu. Przegl Telekomunikacyjny - Wiad Telekomunikacyjne 6(2014):424–427

Waraksa M, Żurek J (2015) VHF Data Exchange System—zintegrowany system wymiany danych w morskim paśmie VHF. Przegl Telekomunikacyjny - Wiad Telekomunikacyjne 4(2015):384–387. https://doi.org/10.15199/59.2015.4.69

Accurate Image Depth Determination for Autonomous Vehicle Navigation

Jurek Z. Sasiadek and Mark J. Walker

List of Abbreviations

CI Confidence Interval
GPS Global Positioning System
INU Inertial Navigation Unit
MAV Mini- or micro-UAV
SVD Singular Value Decomposition
UAV Unmanned Aerial Vehicle

1 Introduction

The authors' research interests encompass the accurate calculation of pose—position and attitude—of an Unmanned Aerial Vehicle (UAV) in an urban environment. The accurate determination of this 6-tuple (x, y, z, pitch, roll, yaw) is important for UAV navigation. Accurate navigation is crucial, for example, to autonomous aerial refueling, a class of problem that has recently received attention in Fravolini et al. (2006), Valasek et al. (2005), and Webb et al. (2007). It should also be noted that accurate navigation is vital to space-based robotic vehicles used for space debris clearance operations. It has been discussed in Kaiser et al. (2006, 2007) that the Global Positioning System (GPS) is the most widely used sensor for aircraft navigation. These papers have pointed out that GPS signals can be blocked. As well, the Volpe Report (Volpe 2001) has reported that GPS signals can be blocked both intentionally and unintentionally, by the leaves of a tree, for example, or a building. It is reasonable to enquire whether other sensors might be used in place of, or as an enhancement to GPS.

J. Z. Sasiadek · M. J. Walker (✉)
Department of Mechanical and Aerospace Engineering, Carleton University, Ottawa, ON, Canada
e-mail: mark_walker@carleton.ca

J. Sasiadek, *Aerospace Robotics III*, GeoPlanet: Earth and Planetary Sciences,
https://doi.org/10.1007/978-3-319-94517-0_5

The pose of a UAV can be estimated using an Inertial Navigation Unit (INU). However, linear position is derived through a double integration of the linear acceleration which is an output of these devices. Consequently, small errors in acceleration quickly result in large cumulative position errors. In addition, if the vehicle undergoes small accelerations, then small errors will swamp the small accelerations, leading to large errors in position. Attitude (pitch, roll, yaw) can be reliably obtained from an INU and altitude (z) can be derived from pressure sensors. Thus, position in the (x, y) plane still needs to be estimated. It is common for UAVs and mini- and micro-UAVs (MAVs) to be equipped with cameras. Can these be used to estimate position in the (xy) plane?

UAV navigation through machine vision has been studied in Valasek et al. (2005), Webb et al. (2007), Kaiser et al. (2006, 2007), Bachrach et al. (2012), Fraundorfer and Scaramuzza (2012), Kanade et al. (2004), and Nister et al. (2006). It is assumed that two images of the environment through which the UAV is flying have been taken. A projective transformation called a homography is defined that maps the pixels of the first-in-time image into the pixels of the second-in-time image. Deriving this homography is a three-step process. In the first, points-of-interest or features are extracted from each image (Torr and Zisserman 2000). In the second step, corresponding points-of-interest or features are found across the two images (Hartley and Zisserman 2003). Finally, a homography is defined from the corresponding points-of-interest (Hartley and Zisserman 2003). Rotation and translation can be derived from this homography if image depth is known (Hu et al. 2009; Michaelson et al. 2004). The process of estimating depth has been discussed in Jaehne (2001) and Iocchi (unavailable). A more robust technique is to rectify the images (Slonka et al. 2008; Loop and Zhang 1999) first and then to estimate depth (Slonka et al. 2008; Hartley and Zisserman 2003).

In this paper, synthetic pixel data is created for which the depth is known to a high degree of accuracy. At this point, the goal is not to generate a definitive depth performance study, rather it is to gain an understanding of the kind of depth performance which might be possible for a small UAV flying close in and far away, at depths of 2 and 20 m, for example. The goal of the research is to report under highly controlled conditions the exact (within a statistical confidence interval) error achieved with software that implements the theory explained below.

The remainder of this paper is organized as described. Section 2 explains the theory of depth calculation using a stereo camera. Section 3 explains how test pixels are generated. Section 4 presents the results and Sect. 5 is the conclusions.

2 Estimating Feature Depth

The authors have published several papers (Sasiadek and Walker 2008, 2010; Sasiadek et al. 2010, 2011) on the problem of accurately finding corners in images and then matching these across images-in-time. Depth of an image point and camera pose is dependent upon finding these matches and has been examined in

Walker and Sasiadek (2013). This paper will concentrate on depth determination and will not consider further the question of finding corners and matching them across images.

In the theory that follows, it is assumed that two consecutive-in-time images I and I′, with feature pixels $x_i = [u_i\ v_i\ 1]^T$ from image I and their matching feature pixels $x'_i = [u'_i\ v'_i\ 1]^T$ from image I′, have been found. Depth calculation proceeds in three steps. First the fundamental matrix is estimated between I and I′, then the two images are rectified. Finally, the depth may be estimated.

2.1 Estimating the Fundamental Matrix

The fundamental matrix, F, between images I and I′ satisfies the equation below

$$x_i'^T F x_i = 0$$
$$F = \begin{bmatrix} f_{11} & f_{12} & f_{13} \\ f_{21} & f_{22} & f_{23} \\ f_{31} & f_{32} & f_{33} \end{bmatrix} \tag{2.1.1}$$

for any pair of matched pixels, x_i and x'_i, $1 \le i \le n$ from images I and I′, respectively. When there are at least 8 of these matched pairs, F can be estimated as described in Chap. 11 of Hartley and Zisserman (2003). Equation (2.1.1) is expanded

$$Af = 0$$
$$\begin{bmatrix} u'_1u_1 & u'_1v_1 & u'_1 & v'_1u_1 & v'_1v_1 & v'_1 & u_1 & v_1 & 1 \\ . & . & . & . & . & . & . & . & . \\ . & . & . & . & . & . & . & . & . \\ . & . & . & . & . & . & . & . & . \\ u'_nu_n & u'_nv_n & u'_n & v'_nu_n & v'_nv_n & v'_n & u_n & v_n & 1 \end{bmatrix} f = 0 \tag{2.1.2}$$
$$f = [f_{11}\ \ f_{12}\ \ f_{13}\ \ f_{21}\ \ f_{22}\ \ f_{23}\ \ f_{31}\ \ f_{32}\ \ f_{33}]$$

for $n \ge 8$ pixel matches across images I and I′ and the row vector f can be found by singular value decomposition (SVD). The value of the matrix F is found by re-arranging f in accordance with Eq. (2.1.1). There are sources of error in solving for f, including pixel location errors and numerical errors caused by the spread of pixel location values around their centre. As well, the third row of F is a linear combination of the first two rows. The fundamental matrix is, thus, of rank 2, while the SVD will produce a solution, in general, of rank 3. These matters are considered and solved in algorithms 11.1 and 11.2 of Chap. 11 of Hartley and Zisserman (2003) and will not be addressed further in this paper.

2.2 Rectifying Images

With an accurate estimate of F, rectifying homographies may now be calculated for the first-in-time image, $H_L = H_{LS}H_{L2}H_{L1}$, and the second-in-time image, H_R $H_L = H_{RS}H_{R2}H_{R1}$, as described in Algorithm 11.1 of Chap. 11 of Slonka et al. (2008). The application of the homographies, $H_{L2}H_{L1}$ and $H_{R2}H_{R1}$, described in detail in the first two steps of Algorithm 11.1 to the pixels x_i and the pixels x'_i, respectively, result in images which are rectified but may be severely distorted (Loop and Zhang 1999). A shearing transform

$$S = \begin{bmatrix} p & q & 0 \\ 0 & 1 & 0 \\ 0 & 0 & 1 \end{bmatrix} \tag{2.2.1}$$

needs to be added on the left, H_{LS}, and another one on the right, H_{RS}. For the purposes of explanation, the remaining discussion will be for the calculation of H_{LS} and H_{RS} is calculated in a similar manner.

Figure 1 depicts an image, with points a, b, c, and lines between them, x and y. The image has width, w, and height, h. Shearing Transform S is to be chosen so that perpendicularity is preserved, Eq. (2.2.2), and aspect ratio is preserved, Eq. (2.2.3).

$$(Sx)^T(Sy) = 0 \tag{2.2.2}$$

$$\frac{(Sx)^T(Sx)}{(Sy)^T(Sy)} = \frac{w^2}{h^2} \tag{2.2.3}$$

Now the points a, b, c, and d must be transformed by $H_{L2}.H_{L1}$ to produce a_1, b_1, c_1, and d_1 respectively. Each of these points is scaled so that its weight is 1. Now the x and y are defined so that

$$\begin{aligned} x &= b_1 - d_1 + 1 = (x_u, x_v, 0) \\ y &= c_1 - a_1 + 1 = (y_u, y_v, 0) \end{aligned} \tag{2.2.4}$$

For Eqs. (2.2.2) and (2.2.3) to be satisfied as closely as possible

$$\begin{aligned} p &= \frac{h^2x_v^2 + w^2y_v^2}{hw(x_vy_u - x_uy_v)} \\ q &= \frac{h^2x_ux_v + w^2y_uy_v}{hw(x_vy_u - x_uy_v)} \end{aligned} \tag{2.2.5}$$

up to sign. The sign for p and q is chosen so that perpendicularity and aspect ratio are minimized. Finally, to maintain image size, a scale factor is introduced (Loop and Zhang 1999). Finally, $H_L = H_{LS}.H_{L2}.H_{L1}$ and $H_R = H_{RS}.H_{R2}H_{R1}$.

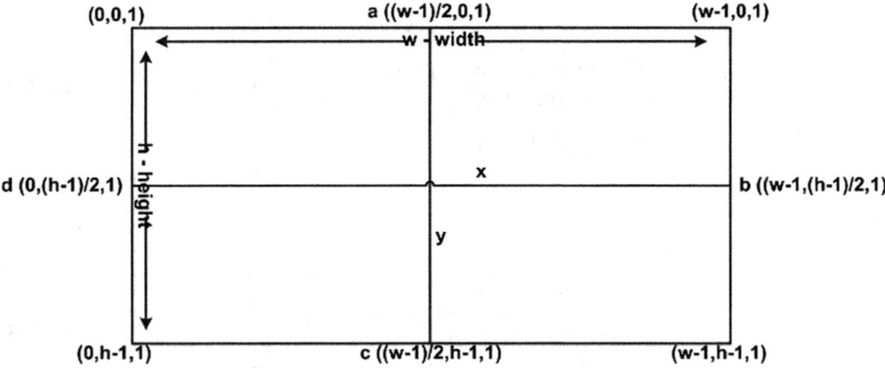

Fig. 1 An image

2.3 *Estimating Image Depth*

The two rectifying homographies, H_L and H_R, are applied, respectively, to the pixels $x_i = [u_i\ v_i\ 1]^T$ from image I and their matching feature pixels $x'_i = [u'_i\ v'_i\ 1]^T$ from image I′,

$$\begin{aligned} \hat{x}_i &= H_L x_i \\ \hat{x}'_i &= H_R x'_i \end{aligned} \tag{2.3.1}$$

and the fundamental matrix F_1 is estimated. Note that if rectification is not to be employed then H_L and H_R are replaced by the Identity matrix. The rectified pixels may contain errors; they can be corrected as described in algorithm 12.1 (Hartley and Zisserman 2003) and the highlights explained further below.

For each pixel pair, translations are defined that take the pixels to the origin.

$$\begin{aligned} T &= \begin{bmatrix} 1 & 0 & -\hat{u}_i \\ 0 & 1 & -\hat{v}_i \\ 0 & 0 & 1 \end{bmatrix} \\ T' &= \begin{bmatrix} 1 & 0 & -\hat{u}'_i \\ 0 & 1 & -\hat{v}'_i \\ 0 & 0 & 1 \end{bmatrix} \end{aligned} \tag{2.3.2}$$

Now

$$F_2 = T'^{-T} F_1 T^{-1} \tag{2.3.3}$$

and the left, $e_L = [e_{L1}\ e_{L2}\ e_{L3}]^T$, and right, $e_R = [e_{R1}\ e_{R2}\ e_{R3}]^T$, epipoles are calculated. Now

$$R = \frac{1}{\sqrt{\left[e_{L1}^2 + e_{L2}^2\right]}} \begin{bmatrix} e_{L1} & e_{L2} & 0 \\ -e_{L2} & e_{L1} & 0 \\ 0 & 0 & 1 \end{bmatrix} \tag{2.3.4}$$
$$\left[e_{L1}^2 + e_{L2}^2\right] \neq 0$$

and

$$R' = \frac{1}{\sqrt{\left[e_{R1}^2 + e_{R2}^2\right]}} \begin{bmatrix} e_{R1} & e_{R2} & 0 \\ -e_{R2} & e_{R1} & 0 \\ 0 & 0 & 1 \end{bmatrix} \tag{2.3.5}$$
$$\left[e_{R1}^2 + e_{R2}^2\right] \neq 0$$

Now

$$\hat{F}_2 = R' F_2 R^T \tag{2.3.6}$$

and

$$\begin{aligned} f &= e_{L3} \\ f' &= e_{R3} \\ a &= \hat{F}_2[2,2] \\ b &= \hat{F}_2[2,3] \\ c &= \hat{F}_2[3,2] \\ d &= \hat{F}_2[3,3] \end{aligned} \tag{2.3.7}$$

With these assignments, g(t) = 0, in Eq. (12.7) of Hartley and Zisserman (2003), can be solved for its 6 real roots, t. It is now necessary to select among 7 values of the variable t, the 6 roots plus t = ∞, the one which minimizes the cost function, Eq. (12.5) of Hartley and Zisserman (2003). The improved pixel values, x_i and x'_i are derived as in algorithm 12.1, based on this minimal value of the variable t. Now

$$\begin{aligned} \hat{x}_i &= T^{-1} R^T x_i \\ \hat{x}'_i &= T'^{-1} R'^T x'_i \end{aligned} \tag{2.3.8}$$

The camera matrices, C_L and C_R, are now obtained. The rotation and translation of the left camera in camera coordinates, R_L and T_L, is assumed known with some, hopefully, small error, as well as the right camera, R_R and T_R. The camera calibration matrix, K, as assumed known and is applicable to the left and right cameras.

Then

$$\begin{aligned} C_L &= H_L K R_L [I \quad 0] \\ C_R &= H_R K R_R [I \quad -(T_R - T_L)] \end{aligned} \tag{2.3.9}$$

Triangulation is now carried out to get the 3D point for each matched and rectified pair of feature pixels, as described in Chap. 12 of Hartley and Zisserman (2003). For a given pair of matched and rectified pixels and assuming the superscript j denotes the j'th row of C_L and C_R and assuming $X = [x\ y\ z\ w]^T$ is the 3D point in homogeneous co-ordinates, then

$$\begin{aligned} AX &= 0 \\ A &= \begin{bmatrix} \hat{u}_i C_L^3 - C_L^1 \\ \hat{v}_i C_L^3 - C_L^2 \\ \hat{u}'_i C_R^3 - C_R^1 \\ \hat{v}_i C_R^3 - C_R^2 \end{bmatrix} \end{aligned} \tag{2.3.10}$$

A solution for X can be found using SVD and depth returned as X[3]/X[4].

2.4 Estimating the Camera Calibration Matrix

The Camera calibration matrix, K, describes the transformation of the location of a 3D scene point into a 2D image point under a specific imaging geometry, unique for a given camera and its photographic settings. Camera calibration produces intrinsic and extrinsic parameters (Bouguet 2015) and has received much attention in the literature (Heikkila 2000; Strat 1984; Tsai 1987; Zhang 2000). The toolbox of J. Bouguet at Caltech (Bouguet 2015) has had multiple releases and it was decided to use this toolbox. The intrinsic camera parameters produced by this toolbox are listed below.

- Focal length in pixels, stored in 2×1 vector **fc**
- Principal point in pixels, stored in 2×1 vector **cc**
- Skew coefficient, stored in scalar **alpha_c**
- Image radial and tangential distortion coefficients, stored in the 5×1 vector **kc**

Using the conventions of Bouguet (2015), the camera matrix is defined as:

$$K = \begin{bmatrix} fc(1) & alpha\_c^* fc(1) & cc(1) \\ 0 & fc(2) & cc(2) \\ 0 & 0 & 1 \end{bmatrix} \tag{2.4.1}$$

3 Test Pixels

Figure 2 depicts the two co-ordinate systems relevant to the reported research. In the material that follows superscripts denote the co-ordinate system the value is relative to—E for Earth, C for camera—while subscripts denote the quantity, as explained. It is assumed that the left and right cameras, C_L and C_R respectively, have the same camera matrix, K. The rotation matrices R_{EC} and R_{CE} rotate earth-to-camera and camera-to-earth co-ordinates, respectively, and follow easily from Fig. 2.

$$\begin{aligned} R_{EC} &= \begin{pmatrix} 0 & 1 & 0 \\ 1 & 0 & 0 \\ 0 & 0 & -1 \end{pmatrix} \\ R_{CE} &= \begin{pmatrix} 0 & 1 & 0 \\ 1 & 0 & 0 \\ 0 & 0 & -1 \end{pmatrix} \end{aligned} \tag{3.1}$$

If the camera co-ordinate system is translated from the Earth co-ordinate system by t_C^E for a camera C, then

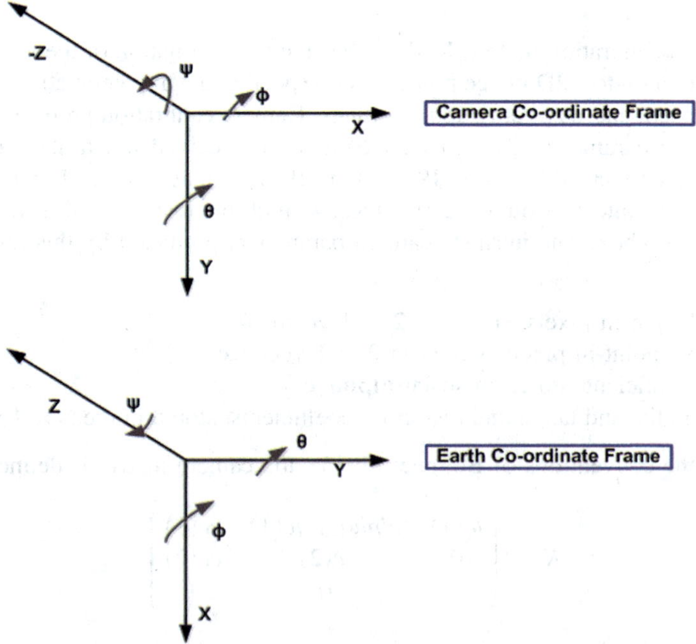

Fig. 2 Co-ordinate systems

$$\begin{aligned} x_i &= KR^C_{C_L} R_{EC}\left(X_i^E - t^E_{C_L}\right) \\ x'_i &= KR^C_{C_R} R_{EC}\left(X_i^E - t^E_{C_R}\right) \end{aligned} \tag{3.2}$$

The camera matrix K comes from camera calibration and has a value in the reported results given below.

$$k = \begin{pmatrix} 6178.88 & 0 & 1372.57 \\ 0 & 6133.65 & 1071.35 \\ 0 & 0 & 1 \end{pmatrix} \tag{3.3}$$

If either camera is rotated about the x-axis, by a_x radians, about the y-axis, by a_y radians, and about the z-axis, by a_z radians, in the camera co-ordinate system, then the rotation matrices R^C_x, R^C_y, and R^C_z follow easily from Fig. 2.

$$\begin{aligned} R^C &= R^C_z . R^C_y . R^C_x \\ R^C_z &= \begin{pmatrix} \cos(a_z) & \sin(a_z) & 0 \\ -\sin(a_z) & \cos(a_z) & 0 \\ 0 & 0 & 1 \end{pmatrix} \\ R^C_y &= \begin{pmatrix} \cos(a_y) & 0 & -\sin(a_y) \\ 0 & 1 & 0 \\ \sin(a_y) & 0 & \cos(a_y) \end{pmatrix} \\ R^C_x &= \begin{pmatrix} 1 & 0 & 0 \\ 0 & \cos(a_x) & \sin(a_x) \\ 0 & -\sin(a_x) & \cos(a_x) \end{pmatrix} \end{aligned} \tag{3.4}$$

For each test case, n = 20 3D points in the Earth co-ordinate system, $X^E_i = (x_i\ y_i\ z_i)^T$, $1 \leq n \leq 20$, are generated in a quadrangle in the xy-plane with height slightly varying—for algorithmic reasons—as

$$\begin{aligned} x_i &= h_1 + (h_2 - h_1) * rand \\ y_i &= w_1 + (w_2 - w_1) * rand \\ z_i &= -d_1 - (d_2 - d_1) * rand \end{aligned} \tag{3.5}$$

where the w and the h values define the width and height of the xy-plane, the d values define the height, and rand is a random number generator producing uniformly random numbers in the range $0 \leq$ rand ≤ 1. The mean value of z_i is $-(d_2 + d_1)/2$ in Earth co-ordinates.

The Eq. (3.2) represents generation of image pixels. However, feature detection performance must follow Sasiadek and Walker (2008, 2010) and adds ±1 pixel Gaussian feature location noise. In this research this error is simulated with a

Gaussian process with mean equal to the pixel position without noise and $\sigma = 3.291$ so that 99.9% of the noise will fall within ± 1 pixel of the mean.

Left and right camera image pixels are processed to yield an estimated depth for each. An average depth can be calculated for the pair of images, an average for the test case can be easily calculated and, using Mendenhall (1968), a confidence interval can be stated for image depth.

In Mendenhall (1968), the objective of statistics is to make inferences about a population based upon sample data. In the case of this research, uniformly distributed pixels are created and then processed in a sometimes linear and sometimes non-linear complex manner to produce estimated depths. The Central Limit Theorem implies that depth will be normally distributed.

For a 95% Confidence Interval (CI) $\alpha = 0.05$. When the sample size, n, is small, one may use Student's t distribution to construct CIs for the mean of a normal distribution (Mendenhall 1968). Thus for image depth, d, the 95% CI for the mean is

$$\bar{d} \pm t_{0.025} \frac{s}{\sqrt{n}} \tag{3.6}$$

It seems reasonable to require $t_{0.025} \leq \sqrt{n}$ and this occur for $n \geq 7$. At $n = 7$, $t_{0.025} = 2.447$, so we need 7 sample depths per test case.

4 Results

In the reported research, the following four basic test cases are run, differentiated in terms of the degree of camera placement error and whether image rectification is carried out, including knowledge of camera rotation. Optimization of pixels prior to depth estimation was not used. The placement error is based on a, hopefully, reasonable estimate of achievable placement error. All data is relative to the camera co-ordinate system. Expected image depth should be −2.0000 or −20.0000 m in the Earth co-ordinate system, as shown below. These depths are to be understood as representative depths at which a small UAV might fly. The goal is to gain an understanding of the kind of performance which might be expected, rather than to conduct a definitive study of depth performance. Actual results are given in Table 1. Figures 3, 4, 5, 6, 7, 8, 9 and 10 present the same results graphically to aid in understanding. The table and associated figures report results without Gaussian pixel error and results with Gaussian pixel error. This latter case is the more accurate representation of the real scenario when corners are located with some small error due, either to imperfections in the pixel formation process or to algorithmic errors.

1. x, y, z translation error both cameras: no error; x, y, z rotation both cameras: no error. No image rectification. No knowledge of camera rotation.

Table 1 Actual image depths

Test number	Sub-test number	No pixel error		Gaussian pixel error	
		Image depth (m) 95% CI	Error image depth (m) 95% CI	Image depth (m) 95% CI	Error image depth (m) 95% CI
1	1	2.0001 ± 0.0003	−0.0001 ± 0.0003	2.0001 ± 0.0004	0.0001 ± 0.0004
1	2	1.4810 ± 0.0001	0.5190 ± 0.0001	1.4810 ± 0.0003	0.5190 ± 0.0003
1	3	0.7239 ± 0.0000	1.2761 ± 0.0000	0.7239 ± 0.0000	1.2761 ± 0.0000
1	4	0.4378 ± 0.0000	1.5622 ± 0.0000	0.4378 ± 0.0000	1.5622 ± 0.0000
1	5	20.0001 ± 0.0004	−0.0001 ± 0.0004	20.0033 ± 0.0082	−0.0033 ± 0.0082
1	6	11.7677 ± 0.0007	8.2323 ± 0.0007	11.7681 ± 0.0013	8.2319 ± 0.0013
1	7	4.4323 ± 0.0007	15.5677 ± 0.0007	4.4322 ± 0.0008	15.5678 ± 0.0008
1	8	2.4705 ± 0.0003	17.5295 ± 0.0003	2.4706 ± 0.0004	17.5294 ± 0.0004
2	1	1.9999 ± 0.0003	0.0002 ± 0.0003	1.9995 ± 0.0006	0.0005 ± 0.0006
2	2	1.9999 ± 0.0004	0.0001 ± 0.0004	2.0000 ± 0.0003	0.0000 ± 0.0003
2	3	1.9999 ± 0.0002	0.0001 ± 0.0002	1.9998 ± 0.0005	0.0002 ± 0.0005
2	4	2.0001 ± 0.0004	−0.0001 ± 0.0004	2.0003 ± 0.0007	−0.0003 ± 0.0007
2	5	19.9999 ± 0.0002	0.0001 ± 0.0002	19.9979 ± 0.0103	0.0021 ± 0.0103
2	6	20.0001 ± 0.0002	−0.0001 ± 0.0002	19.9954 ± 0.0093	0.0046 ± 0.0093
2	7	20.0000 ± 0.0005	−0.0000 ± 0.0005	20.0022 ± 0.0058	−0.0022 ± 0.0058
2	8	20.0001 ± 0.0004	−0.0001 ± 0.0004	19.9946 ± 0.0110	0.0054 ± 0.0110
3	1	2.0706 ± 0.0005	−0.0706 ± 0.0005	2.0745 ± 0.0090	−0.0745 ± 0.0090
3	2	2.0707 ± 0.0004	−0.0707 ± 0.0004	2.0764 ± 0.0040	−0.0764 ± 0.0040
3	3	2.0710 ± 0.0003	−0.0710 ± 0.0003	2.0821 ± 0.0041	−0.0821 ± 0.0041
3	4	2.0710 ± 0.0002	−0.0710 ± 0.0002	2.0756 ± 0.0114	−0.0756 ± 0.0114
3	5	21.4985 ± 0.0058	−1.4985 ± 0.0058	20.6108 ± 0.1846	−0.6108 ± 0.1846

(continued)

Table 1 (continued)

Test number	Sub-test number	No pixel error		Gaussian pixel error	
		Image depth (m) 95% CI	Error image depth (m) 95% CI	Image depth (m) 95% CI	Error image depth (m) 95% CI
3	6	21.4914 ± 0.0185	−1.4914 ± 0.0185	20.1190 ± 0.4369	−0.1190 ± 0.4369
3	7	21.4715 ± 0.0103	−1.4715 ± 0.0103	20.8610 ± 0.4382	−0.8610 ± 0.4382
3	8	21.4583 ± 0.0108	−1.4583 ± 0.0108	20.4718 ± 0.3235	−0.4718 ± 0.3235
4	1	1.9334 ± 0.0001	0.0666 ± 0.0001	1.9272 ± 0.0024	0.0728 ± 0.0024
4	2	1.9334 ± 0.0003	0.0666 ± 0.0003	1.9271 ± 0.0031	0.0729 ± 0.0031
4	3	1.9330 ± 0.0003	0.0670 ± 0.0003	1.9274 ± 0.0045	0.0726 ± 0.0045
4	4	1.9331 ± 0.0002	0.0669 ± 0.0002	1.9239 ± 0.0058	0.0761 ± 0.0058
4	5	18.7049 ± 0.0046	1.2951 ± 0.0046	19.9721 ± 1.0226	0.0279 ± 1.0226
4	6	18.7086 ± 0.0110	1.2914 ± 0.0110	19.4199 ± 0.1835	0.5801 ± 0.1835
4	7	18.7322 ± 0.0112	1.2678 ± 0.0112	20.5253 ± 1.1037	−0.5253 ± 1.1037
4	8	18.7365 ± 0.0088	1.2635 ± 0.0088	19.8085 ± 0.7882	0.1915 ± 07882

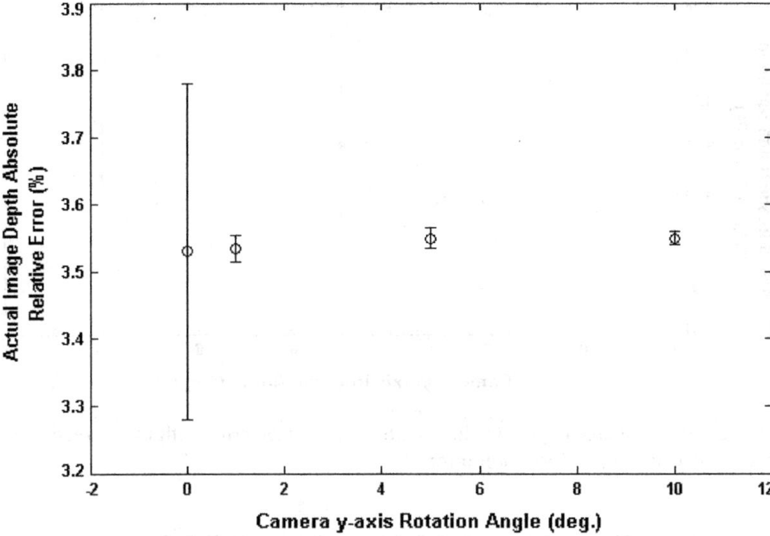

Fig. 3 Expected 2 m image depth absolute relative percentage error with image rectification, test 3, sub-tests 1–4 with no pixel location error

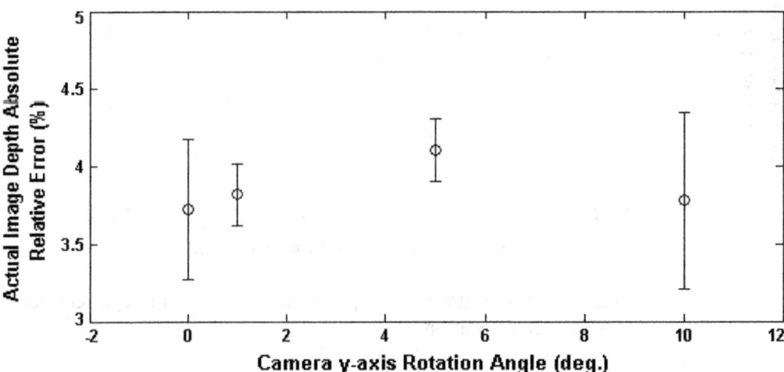

Fig. 4 Expected 2 m image depth absolute relative percentage error with image rectification, test 3, sub-tests 1–4 with Gaussian pixel location error

2. x, y, z translation error both cameras: no error; x, y, z rotation both cameras: no error. Image rectification.
3. x, y, z translation error both cameras: left—0.002 m right— −0.002 m; x, y, z rotation error both cameras: left—0.05° right— −0.05°. Image rectification.

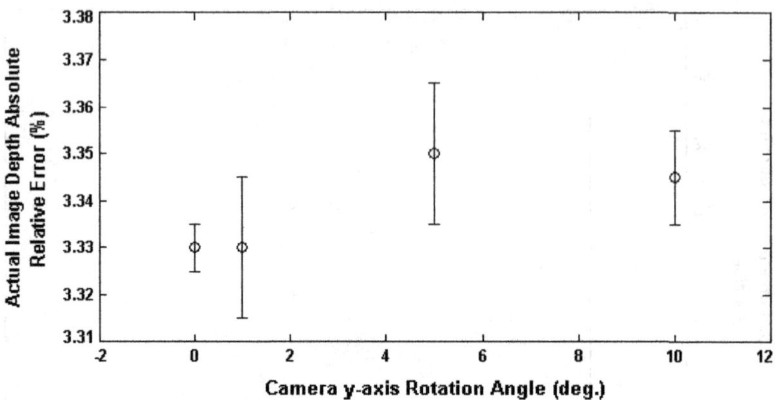

Fig. 5 Expected 2 m image depth absolute relative percentage error with image rectification, test 4, sub-tests 1–4 with no pixel location error

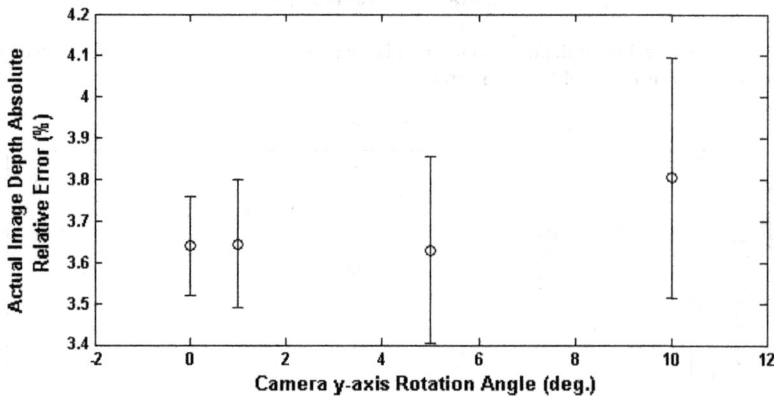

Fig. 6 Expected 2 m image depth absolute relative percentage error with image rectification, test 4, sub-tests 1–4 with Gaussian pixel location error

4. x, y, z translation error both cameras: left— −0.002 m right—0.002 m;
   x, y, z rotation error both cameras: left— −0.05° right—0.05°.
   Image rectification.

Within each of these tests are 8 sub-tests.
For the first four:
Earth points: $w_1 = 0.02$, $w_2 = 0.06$; $h_1 = -0.3$, $h_2 = 0.2$; $d_1 = 1.9975$, $d_2 = 2.0025$ for Eq. (3.5).

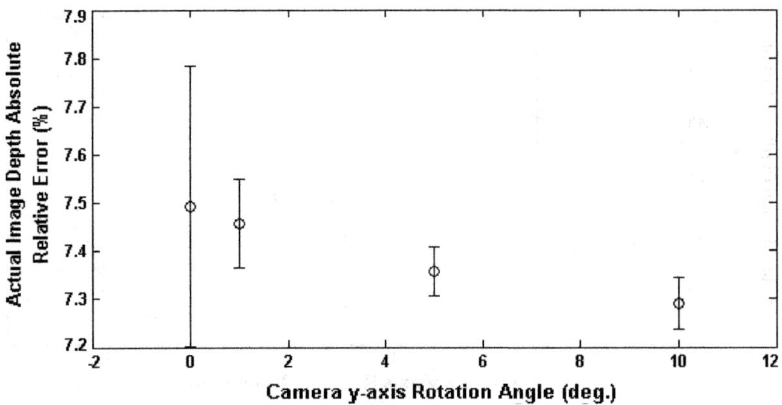

Fig. 7 Expected 20 m image depth absolute relative percentage error with image rectification, test 3, sub-tests 5–8 with no pixel location error

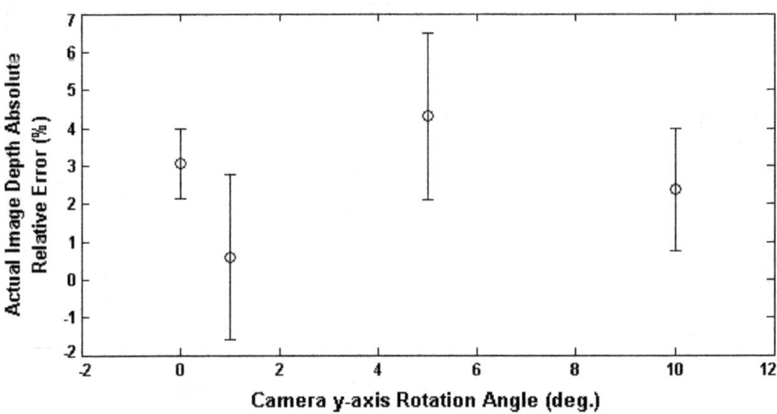

Fig. 8 Expected 20 m image depth absolute relative percentage error with image rectification, test 3, sub-tests 5–8 with Gaussian pixel location error

For the last four:
Earth points: $w_1 = -0.35$, $w_2 = 1.15$; $h_1 = -3.37$, $h_2 = 2.92$; $d_1 = 19.9975$, $d_2 = 20.0025$ for Eq. (3.5).

1. Pose of left camera: (x, y, z): (−0.1 m, 0.0 m, 0.0 m), (0.0°, 0.0°, 0.0°)
   Pose of right camera: (x, y, z): (0.1 m, 0.0 m, 0.0 m), (0.0°, 0.0°, 0.0°)
2. Pose of left camera: (x, y, z): (−0.1 m, 0.0 m, 0.0 m), (0.0°, −1.0°, 0.0°)
   Pose of right camera: (x, y, z): (0.1 m, 0.0 m, 0.0 m), (0.0°, 1.0°, 0.0°)
3. Pose of left camera: (x, y, z): (−0.1 m, 0.0 m, 0.0 m), (0.0°, −5.0°, 0.0°)
   Pose of right camera: (x, y, z): (0.1 m, 0.0 m, 0.0 m), (0.0°, 5.0°, 0.0°)
4. Pose of left camera: (x, y, z): (−0.1 m, 0.0 m, 0.0 m), (0.0°, −10.0°, 0.0°)
   Pose of right camera: (x, y, z): (0.1 m, 0.0 m, 0.0 m), (0.0°, 10.0°, 0.0°)

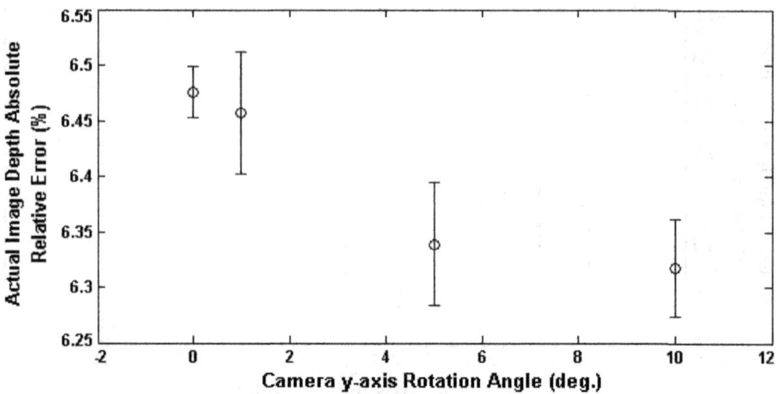

Fig. 9 Expected 20 m image depth absolute relative percentage error with image rectification, test 4, sub-tests 5–8 with no pixel location error

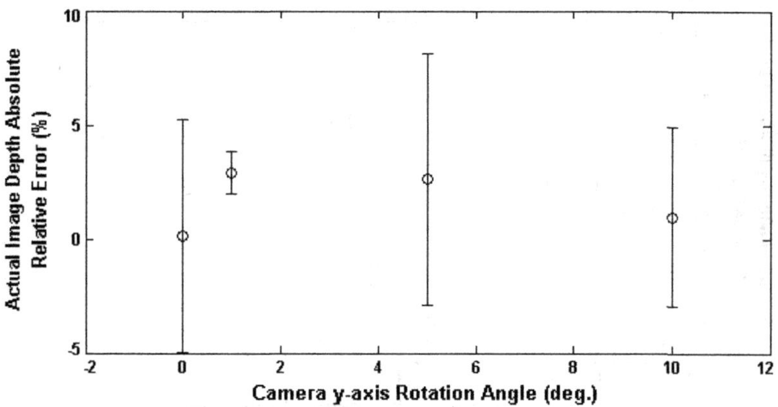

Fig. 10 Expected 20 m image depth absolute relative percentage error with image rectification, test 4, sub-tests 5–8 with Gaussian pixel location error

5. Pose of left camera: (x, y, z): (−0.5 m, 0.0 m, 0.0 m), (0.0°, 0.0°, 0.0°)
   Pose of right camera: (x, y, z): (0.5 m, 0.0 m, 0.0 m), (0.0°, 0.0°, 0.0°)
6. Pose of left camera: (x, y, z): (−0.5 m, 0.0 m, 0.0 m), (0.0°, −1.0°, 0.0°)
   Pose of right camera: (x, y, z): (0.5 m, 0.0 m, 0.0 m), (0.0°, 1.0°, 0.0°)
7. Pose of left camera: (x, y, z): (−0.5 m, 0.0 m, 0.0 m), (0.0°, −5.0°, 0.0°)
   Pose of right camera: (x, y, z): (0.5 m, 0.0 m, 0.0 m), (0.0°, 5.0°, 0.0°)
8. Pose of left camera: (x, y, z): (−0.5 m, 0.0 m, 0.0 m), (0.0°, −10.0°, 0.0°)
   Pose of right camera: (x, y, z): (0.5 m, 0.0 m, 0.0 m), (0.0°, 10.0°, 0.0°)

5 Conclusions

It is the goal of the reported research eventually to accurately predict robotic pose to allow accurate navigation in the absence of Global Positioning System (GPS) or similar signals, using stereo camera-equipped Unmanned Aerial Vehicles (UAVs) or robotic devices in general. Accurate navigation is applied to UAVs in an urban environment, to aerial refueling, and to space-based robotic vehicles engaged in space debris clearance operations. At this point, the goal is not to generate a definitive depth performance study, rather it is to gain an understanding of the kind of depth performance which might be possible for a small UAV flying close in and far away, at depths of 2 and 20 m, for example.

Pose prediction using stereo cameras requires an estimate of the relative pose of the left and right cameras. This knowledge, obviously, will not be available without error. Test 1 assumes no knowledge of camera rotation without the use of image rectification; test 2 still assumes perfect knowledge, but now with image rectification; while test 3 and 4 assume, respectively, positive, and then negative errors in position and angle, with image rectification. Within each test there are 8 sub-tests with various y-axis camera rotations and x, y, z-axis camera translations.

The results are shown in Table 1 and graphically in Figs. 3, 4, 5, 6, 7, 8, 9 and 10.

For the 2 m expected depth case, actual depths, without image rectification and with no knowledge of camera rotation, are acceptable only when the optical axes of the two cameras are parallel, and without any axial rotations. When a camera y-axis rotation of $-1.0°$ is introduced at the left camera, and $+1.0°$ is introduced at the right camera a 26% relative image depth error is introduced. Similarly, for the 20 m. expected depth case, actual depths, without image rectification and with no knowledge of camera rotation, are acceptable only when the optical axes of the two cameras are parallel and without any axial rotations. When a camera y-axis rotation of $-1.0°$ is introduced at the left camera, and $+1.0°$ is introduced at the right camera a 41% relative image depth error is introduced. Worse performance is shown when the y-axis rotation is increased to 5.0° and then 10.0°. There is very little effect produced by assuming a Gaussian pixel error. This is so because the 3σ error is ± 1 pixel, so that the actual error in any given test run will be much less than ± 1 pixel.

With image rectification and over a y-axis rotation of 0.0°, 1.0°, 5.0°, and 10.0°, actual image depth is nearly a perfect match to expected image depth when no pose error is assumed. More realistically, there will be a pose error and this can introduce an acceptably small error in depth as shown in Table 1.

Table 1, as well as Figs. 3, 4, 5, 6, 7, 8, 9 and 10 show the results with no pixel position error, or even Gaussian pixel position error, and clearly shows with image rectification and for the stated camera pose errors and up to 10° camera y-axis rotation, image depth can be known with less than 4.3% relative error when the expected image depth is 2 m. When expected image depth is 20 m and no pixel error is assumed then actual image depth can be known with less than 7.6% relative error. Relative error increases to a maximum of 8.2% when Gaussian pixel error is assumed for a 20-m expected image depth.

In general, best performance is achieved with small optical axis y-axis rotations on the left and right camera.

The image depth chosen was restrictive, only 2 and 20 m performance was studied. Future research should look at more and greater image depths, say between 2 and 100 m. Also, image depth performance without image rectification should be examined for camera pose angles of less than 5°, say 0.5°, 1.0°, 2.0°, and 4.0°. It is possible that small pose errors can still give reasonable performance without using image rectification.

References

Bachrach A, Prentice S, He R, Henry P, Huang A, Krainin M, Maturana D, Fox D, Roy N (2012) Estimation, planning and mapping for autonomous flight using an RGB-D camera in GPS-denied environments. Int J Robot Res 31(11):1320–1343. https://doi.org/10.1177/0278364912455256

Bouguet J (2015) Camera calibration toolbox for matlab. http://www.vision.caltech.edu/bouguetj/calib_doc. Accessed 18 Dec 2009

Fraundorfer F, Scaramuzza D (2012) Visual odometry part II: matching, robustness, optimization, and applications. IEEE Robot Autom Mag 19(2):78–90

Fravolini M, Campa G, Ficola A, Napolitano M, Seanor B (2006) Modeling and control issues for machine vision-based autonomous aerial refueling for UAVs. J Guid Control Dyn (unpublished)

Hartley R, Zisserman A (2003) Multiple view geometry in computer vision. Cambridge University Press, Cambridge

Heikkila J (2000) Geometric camera calibration using circular control points. IEEE Trans Pattern Anal Mach Intell 22(10):1066–1077

Hu G, MacKunis W, Gans N, Dixon W, Chen J, Behal A, Dawson D (2009) Homography-based visual servo control with imperfect camera calibration. IEEE Trans Autom Control 54 (6):1318–1324

Iocchi L (unavailable) Stereo vision: triangulation. http://www.dis.uniroma1.it/~iocchi/stereo/triang.html. Accessed 27 Jan 2011

Jaehne B (2001) 3-D imaging. In: Jaehne B (ed) Digital image processing, 5th edn. Springer, Berlin

Kaiser K, Gans N, Mehta S, Dixon W (2006) Position and orientation of an aerial vehicle through chained, vision-based pose reconstruction. In: Proceedings of the AIAA guidance, navigation, and control conference and exhibit. AIAA-02006-6719

Kaiser K, Gans N, Dixon W (2007) Localization and control of an aerial vehicle through chained, vision-based pose reconstruction. In: IEEE proceedings of the 2007 American control conference, pp 5934–5939

Kanade T, Amidi O, Ke Q (2004) Real-time and 3D vision for autonomous small and micro air vehicles. In: Proceedings of the 43rd IEEE conference on decision and control, pp 1655–1662

Loop C, Zhang Z (1999) Computing rectifying homographies for stereo vision. Proc IEEE Conf Comput Vis Pattern Recognit 1:125–131

Mendenhall W (1968) Introduction to linear models and the design and analysis of experiments. Wadsworth Publishing Company, Belmont, California

Michaelson E, Kirchhof M, Stilla U (2004) Sensor pose inference from airborne videos by decomposing homography estimates. In: Proceedings of the XXth ISPRS congress, Technical Commission III

Nister D, Naroditsky O, Bergen J (2006) Visual odometry for ground vehicle applications. J Field Robot 23(1):3–20

Sasiadek J, Walker M (2008) Vision-based UAV navigation. In: Proceedings of the AIAA guidance, navigation, and control conference and exhibit. AIAA 2008-6667

Sasiadek J, Walker M (2010) Feature detector performance for UAV navigation. In: Proceedings of IASTED modelling, identification, and control conference. https://doi.org/10.2316/p.2010.675-114

Sasiadek J, Walker M, Krzyzak A (2010) Feature matching for UAV navigation in urban environments. In: Proceedings of conference methods and models in automation and robotics. https://doi.org/10.1109/mmar.2010.5587244

Sasiadek J, Walker M, Krzyzak A (2011) Accurate feature matching for autonomous vehicle navigation in urban environments. In: Proceedings of conference methods and models in automation and robotics. https://doi.org/10.1109/mmar.2011.6031318

Slonka M, Hlavac V, Boyle R (2008) Image processing, analysis, and machine vision. Thomson, USA

Strat T (1984) Recovering the camera parameters from a transformation matrix. In: Proceedings of DARPA image understanding workshop, pp 264–271

Torr P, Zisserman A (2000) Feature based methods for structure and motion estimation. In: Vision algorithms: theory and practice: international workshop on vision algorithms, pp 278–294

Tsai R (1987) A versatile camera calibration technique for high-accuracy 3D machine vision metrology using off-the-shelf TV cameras and lenses. IEEE J Robot Autom RA-3(4):323–344

Valasek J, Gunnam K, Kimmet J, Tandale M, Junkins J, Hughes D (2005) Vision based sensor and navigation system for autonomous air refueling. J Guid Control Dyn 28(5):979–989

Volpe JA (2001) N.T.S. Center, Vulnerability assessment of the transport infrastructure relying on the global positioning system. Report, Office of the Assistant Secretary for Transportation Policy, U.S. Department of Transportation

Walker M, Sasiadek J (2013) Accurate pose determination for autonomous vehicle navigation. In: Proceedings of conference methods and models in automation and robotics. https://doi.org/10.1109/mmar.2013.6669933

Webb T, Prazenica R, Kurdial A, Lind R (2007) Vision-based state estimation for autonomous micro air vehicles. J Guid Control Dyn 30(3):816–826

Zhang Z (2000) A flexible new technique for camera calibration. IEEE Trans Pattern Anal Mach Intell 22(11):1330–1334

Pose Estimation for Mobile and Flying Robots via Vision System

Malik M. A. Al-Isawi and Jurek Z. Sasiadek

1 Introduction

A camera has been used as a sensor in many applications related to vision based estimation techniques. The vision based pose estimation technique evaluates and calculates relative translations and rotations of two images for the same target. DeMenthon and Davis (1992) and Quan and Lan (1999) propose methods to reconstruct the camera's pose with respect to an object. In this case, an accurate geometric model of the object is needed. The pose estimation from homography and investigation of measurement errors have been shown in Michaelson et al. (2004). In order to evaluate the robot's motion, the general homography decomposition was used by Ma et al. (2003). Faugeras and Lustman (1988) described a more specific decomposition for the homography when the robot is moving on a planar surface. Both approaches provide two possible solutions for the camera pose. A new algorithm for relative pose estimation between two images based on a new decomposition was proposed in Montijano and Sagues (2009). The Iterative Closest Point (ICP) algorithm was used in Besl and McKay (1992) to estimates position and attitude based on 3D data. The mathematical preliminaries of computing the closest point were shown in Walker and Sasiadek (2013). The performance was achieved with three different focal lengths. A closed-form a solution for three or more points using unit quaternions were presented in Horn (1987). The pose estimation for UAV has been shown Walker and Sasiadek (2015). This algorithm allows accurate navigation when the GPS is absence using a stereo camera in UAV. Xu et al. (2014)

M. M. A. Al-Isawi (✉)
Al-Khwarizmi Engineering. College, University of Baghdad, Baghdad, Iraq
e-mail: malikalisawi@cmail.carleton.ca

J. Z. Sasiadek
Department of Mechanical and Aerospace Engineering,
Carleton University, Ottawa, ON, Canada
e-mail: jurek_sasiadek@carleton.ca

J. Sasiadek, *Aerospace Robotics III*, GeoPlanet: Earth and Planetary Sciences,
https://doi.org/10.1007/978-3-319-94517-0_6

develops a new algorithm based on EKF to Pose estimation for UAV aerial refueling with serious turbulences. A new algorithm for the Minimal number of points Linear pose estimation with known Zenith direction (MLZ) was presented by Kniaz (2016). This algorithm was estimated a pose of a camera with known gravity vector.

There are many methods that can be used to find the pose estimation (i.e. fundamental matrix, and homography matrix). We used homography matrix for two reasons: firstly, the homography method used two different plans to compare points while the fundamental matrix used projection plane for the same points. In addition, the homography method needs four points to determine it while the fundamental method needs at least eight points. In this paper, three different algorithms were used to estimate the homography matrix: (Direct Linear Transformation (DLT), Random Sample Consensus (RANSAC), and Pseudo-Inverse (PINV)). The pose estimation can be found from fast techniques for homography decomposition, Iterative Closest Point Algorithm (ICP) and Horn's absolute orientation method. Simulation results show that these methods are robust to pixel noise. The depth can be found by using the stereo camera for both mobile and flying robots. All results are compared with an actual pose.

2 Theoretical Background

At the start, it is convenient to define the geometry projection model of the camera which is used. Consider a camera with reference frame (O) captures two images at two locations with the same sense in reference (π). The homography can be computed from the relationship between the two locations using the 2D pixel information and the camera model as shown in Fig. (1).

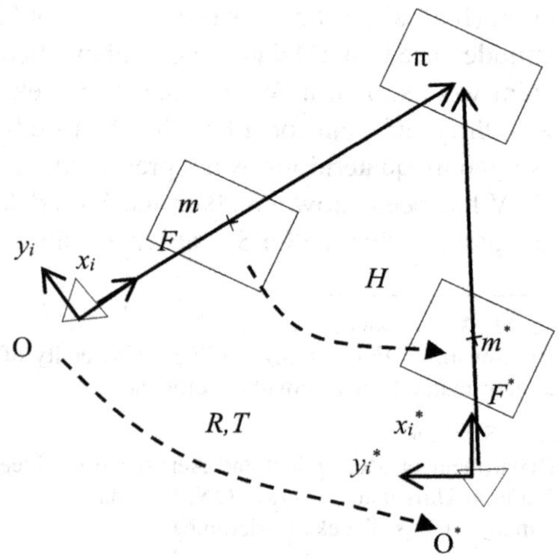

Fig. 1 The geometry of a homography mapping

2.1 Feature Matching

The feature matching process consists of two steps. First, using SURF methods to extract the features from two input images. Then, using some points of interests features can be matched between two images.

2.2 Modeling

We assume the camera is moving in the $x_i\, y_i$ plane while z-axis is orthogonal to $x_i\, y_i$. The camera views four or more planar and non-collinear feature points lying in a plane in front it. These points expressed in the frame (F and F^*) are denoted by $(x_i\; y_i\; z_i)$ and $\left(x_i^*\, y_i^*\, z_i^*\right)$ ϵR respectively as shown by Hu et al. (2009), and the normalized Euclidean coordinate vector can be express by m and m* ϵR^3

$$m = \begin{bmatrix} \frac{x_i}{z_i} \frac{y_i}{z_i} 1 \end{bmatrix}^T \quad m^* = \begin{bmatrix} \frac{x_i^*}{z_i^*} \frac{y_i^*}{z_i^*} 1 \end{bmatrix}^T \tag{2.1}$$

Feature points m and m^* in frame (F and F^*) respectively are related by the homography matrix H ϵR^{3x3} (Faugeras and Lustman 1988) as:

$$m = \alpha_i H m^* \tag{2.2}$$

$$m = \frac{z_i^*}{z_i}\left(R + \frac{t}{d}n^T\right)m^* \tag{2.3}$$

$$H = \left(R + \frac{t}{d}n^T\right) \tag{2.4}$$

where $\alpha_i = \frac{z_i^*}{z_i}$, is called the scalar depth ratio. The H is homography matrix and can be decomposed to R, t/d and n., the R is a rotation matrix, t is the translation vector of the camera, d is the distance between the origin camera to the plane π, and n is the surface normal vector to the plane π as Faugeras and Lustman (1988) and Bouguet (2008).

From projective geometry, the Euclidean coordinates for (m and m^*) can be expressed in image coordinates (pixel coordinates) as

$$P = km \text{ and } p^* = km^* \tag{2.5}$$

where kϵR^{3x3} is a constant camera matrix and can be determine from camera calibration (Xuebo 2008),

$$k = \begin{bmatrix} f_x & \beta & u_o \\ 0 & f_y & v_o \\ 0 & 0 & 1 \end{bmatrix} \tag{2.6}$$

where k is an intrinsic camera matrix,

(f_x, f_y): is the focal length in x and y directions respectively. (u_0 and v_0): are the coordinates of the principal point in terms of pixel dimensions. β is the skew coefficient between the camera's *x*- and *y*-axes, which is often zero. (P and p^*): are pixel coordinates of two images denoted by:

$$P = [u\ v\ 1]^T \ and\ P^* = [u^*\ v^*\ 1] \tag{2.7}$$

Using Eq. (2.5) into relation Eq. (2.2), we get

$$P = \alpha_i kHk^{-1}\ P^* = \alpha_i GP^* \tag{2.8}$$

It is possible to compute the homography matrix from two given images of a planar target by using the DLT (Direct Linear Transfer), PINV (Pseudo-Inverse method) and the RANSAC algorithm applying to classify correspondences as inliers or outliers (Ma et al. 2003). In fact, four non-collinear ($n \geq 4$) matched points suffice to compute G up to a scale factor. Then, using a known matrix K, we compute the matrix H up to a scale factor.

3 Decomposition of Homographies

The main idea of this section is to extract 2-D information from the environment using the camera image and then determine rotation and translation vector from O to O* from homography matrix. According to the planner motion of the camera, the rotation R and translation T can be written as follows:

$$R = \begin{bmatrix} \cos\theta & -\sin\theta & 0 \\ \sin\theta & \cos\theta & 0 \\ 0 & 0 & 1 \end{bmatrix} \tag{3.1}$$

$$T = \frac{t}{d} = \begin{bmatrix} \frac{t_x}{d} & \frac{t_y}{d} & o \end{bmatrix}^T = \begin{bmatrix} T_1 & T_2 & 0 \end{bmatrix}^T \tag{3.2}$$

where θ∊R is the rotational angle around z_i axis between O to O^*, t_x, and t_y the translate along x_i and y_i respectively, and d is the distance from origin of O to the reference plane π. the H matrix can be calculate by substituting (2.8) and (3.1) into (2.4) (Xuebo 2008):

$$H = \begin{bmatrix} \cos(\theta) + T_1 n_1 & -\sin(\theta) + T_1 n_2 & T_1 n_3 \\ \sin(\theta) + T_2 n_1 & \cos(\theta) + T_2 n_2 & T_2 n_3 \\ 0 & 0 & 1 \end{bmatrix} = \begin{bmatrix} h_{11} & h_{12} & h_{13} \\ h_{21} & h_{22} & h_{23} \\ 0 & 0 & 1 \end{bmatrix} \tag{3.3}$$

It's clear from (3.3), the algorithm for homography decomposition will be divided into three cases depending on the h_{13} and h_{23} Xuebo (2008).

Case (1) $h_{13}^2 \neq 0,\ h_{23}^2 \neq 0$

$$T_1 = \varepsilon_1 \sqrt{\frac{h_{13}^2[(h_{12}+h_{21})^2 + (h_{11}-h_{22})^2}{h_{13}^2+h_{23}^2} + h_{13}^2} \tag{3.4}$$

$$T_2 = \varepsilon_1 sgn(h_{23}h_{13}) \sqrt{\frac{h_{23}^2[(h_{12}+h_{21})^2 + (h_{11}-h_{22})^2}{h_{13}^2+h_{23}^2} + h_{23}^2} \tag{3.5}$$

where $\varepsilon_1 = \pm 1$, it can be noted that there are two solutions of T that can be chosen according to Xuebo (2008). Also, we can compute the n vector in (3.3) as follows:

$$n_1 = \frac{t_1(h_{11}-h_{22}) + t_2(h_{12}+h_{21})}{t_1^2+t_2^2} \tag{3.6}$$

$$n_2 = \frac{t_1(h_{12}+h_{21}) - t_2(h_{11}-h_{22})}{t_1^2+t_2^2} \tag{3.7}$$

$$n_3 = \varepsilon_1 sgn(h_{13}) \sqrt{\frac{h_{13}^2+h_{23}^2}{t_1^2+t_2^2}} \tag{3.8}$$

It's easy to compute the rotation matrix from (2.4).
Case (2) $h_{13}^2 \neq 0,\ h_{23}^2 = 0$ or $h_{13}^2 = 0,\ h_{23}^2 \neq 0$

$$T_1 = \varepsilon_1 \sqrt{\frac{h_{13}^2[(h_{12}+h_{21})^2 + (h_{11}-h_{22})^2}{h_{13}^2+h_{23}^2} + h_{13}^2} \quad T_2 = 0 \tag{3.9}$$

$$n_1 = (h_{11}-h_{22})/t_1 \tag{3.10}$$

$$n_2 = (h_{12}+h_{21})/t_1 \tag{3.11}$$

$$n_3 = h_{13}/t_1 \tag{3.12}$$

Also, we can calculate the rotation matrix from (2.4).
Case (3) $h_{13}^2 = 0$ and $h_{23}^2 = 0$

In this case, we have: If $h_{11} = h_{22}$, $h_{21} = -h_{12}$ then T1 = T2 = 0 and the homography degenerates to a rotation matrix.

Otherwise, the decomposition problem degenerates to a simplified version of two by two homography matrix. The decomposition can be done using the singular value decomposition (SVD) techniques similar to the Faugeras' algorithm.

4 Iterative Closest Point Algorithm (ICP)

An ICP algorithm attempts to match two sets of points. One of these sets might be a reference image, while the other is a set of data points describing the ranges to certain points on an object. The ICP algorithm (Besl and McKay 1992; Walker and Sasiadek 2013)provides an estimate of camera translation and rotation between data taken from stereo images. The ICP algorithm is as follows:

A point set P of the source image point with N_p and the destination points become X with N_x are given. The algorithm is following:

1. Iteration is initialized with $P_0 = P,\ \vec{q_0} = [1,\ \ 0,\ \ 0,\ \ 0,\ \ 0,\ \ 0,\ \ 0]^t$
2. Compute the closest points: $Y_k = C\ (P_k,\ X) = \{x_i,\ i = 1,\ldots.N_x\}$.
3. Compute the registration: $(q_k,\ d_k) = Q\ (P_0,\ Y_k)$.
4. Compute the centers of mass for μ_p *and* μ_x

$$\mu_p = \frac{1}{N_p}\sum_{i=1}^{N_p} p_i,\ \mu_x = \frac{1}{N_x}\sum_{i=1}^{N_x} x_i \tag{4.1}$$

5. The cross-covariance is

$$\sum px = \frac{1}{N_p}\sum_{i=1}^{N_p} [(p_i - \mu_p)(x_i - \mu_x)]^T. \tag{4.2}$$

6. The cross-covariance can be put into the form of a matrix.

$$Q\left(\sum_{px}\right) = \begin{bmatrix} tr\left(\sum_{px}\right) & \Delta^T \\ \Delta & \sum_{px} + \sum_{px}^T - tr(\sum_{px})I_3 \end{bmatrix} \tag{4.3}$$

where I_3 is the 3×3 identity matrix,

$$\Delta = [A_{23}, A_{31}, A_{12}] \quad \text{and} \quad A_{ij} = \left(\sum_{px} - \sum_{px}^T\right)_{ij} \tag{4.4}$$

7. Apply the registration: $P_{k+1} = q_R(P_0) + q_T$
   where $q_R = [q_0\ q_1\ q_2\ q_3]^T$ is the unit eigenvector of (4.3), the optimal translation vector is

$$q_T = \mu_x - R(q_R)\mu_p \tag{4.5}$$

8. Terminate the iteration when the change in mean-square error falls below a threshold $\tau > 0$; $dk - dk + 1 < \tau$.
   The mean square objective function is:

$$f(q) = \frac{1}{N_p}\sum_{i=1}^{N_P} \|x_i - R(q_R)p_i - q_T\|^2 \quad (4.6)$$

9. Steps 1–5 are repeated until convergence is within the tolerance.

5 Horn's Absolute Orientation Method

The transformation between two Cartesian coordinate systems can be thought of as the result of a rigid-body motion and can thus be decomposed into a rotation and a translation. Let the coordinates of the four points in each of the two coordinate systems be a $P_{A,i}$ and $P_{B,i}$ for two frames A and B. Horn (1987) present mathematical relationship between these sets of points is:

$$P_{A,i} = sR_{A/B}P_{B,i} + T_{A/B} \quad (5.1)$$

where s is a scale vector and $R_{A/B}$ and $T_{A/B}$ are rotation and translation matrices between frame A and frame B.

The absolute orientation algorithm as follow:

- Subtracting the centroid of the point sets for both frames.

$$\acute{P}_{A,i} = P_{A,i} - \sum_j P_{A,j}, \quad \acute{P}_{B,i} = P_{B,i} - \sum_j P_{B,j} \quad (5.2)$$

- The intermediate matrices

$$M = \begin{bmatrix} S_{xx} & S_{xy} & S_{xz} \\ S_{yx} & S_{yy} & S_{yz} \\ S_{zx} & S_{zy} & S_{zz} \end{bmatrix} = \sum_i \acute{P}_{A,i}\acute{P}_{B,i}^T \quad (5.3)$$

- And

$$N = \begin{bmatrix} S_{xx}+S_{yy}+S_{zz} & S_{yz}-S_{zy} & S_{zx}-S_{xz} & S_{xy}-S_{yx} \\ S_{yz}-S_{zy} & S_{xx}-S_{yy}-S_{zz} & S_{xy}+S_{yx} & S_{zx}+S_{xz} \\ S_{zx}-S_{xz} & S_{xy}+S_{yx} & -S_{xx}+S_{yy}-S_{zz} & S_{yz}+S_{zy} \\ S_{xy}-S_{yx} & S_{zx}+S_{xz} & S_{yz}+S_{zy} & -S_{xx}-S_{yy}+S_{zz} \end{bmatrix} \quad (5.4)$$

- The rotation between frames A and B is the quaternion matrix N which is the eigenvector corresponding to the maximum eigenvalue of N.

$$R_{A/B} = R(q_{A/B}). \tag{5.5}$$

- Then quaternion minimizes by the mean squared error function:

$$f = \sum_i \left\| P_{A,i} - sR_{A/B}P_{B,i} - T_{A/B} \right\|^2 \tag{5.6}$$

- The scale can be found as,

$$s = \frac{\sum_i \left\| \acute{P}_{A,i} \right\|^2}{\sum_i \left\| \acute{P}_{B,i} \right\|^2} \tag{5.7}$$

- The translation vector can be found by solving for the last remaining variable in Eq. (5.1).

6 Experimental Results and Analysis

To verify the performance of all algorithms for pose estimation and find the translation and rotation between two cameras, we should first calculate the intrinsic camera matrix K from the calibration camera Bouguet (2008).

6.1 Camera Calibration

The first step is to determine a relationship between the object that appears on images and where it is in the world coordinate. This process is called the Camera Calibration. Camera calibration is the process of calculating intrinsic parameters such as the focal length, principal point and lens distortion and extrinsic parameters such as the 3D position and orientation of the camera with respect to the world coordinate system. The inputs of this toolbox are several images of a model chessboard plane containing the calibration points. Corners of the calibration chessboard plane are used as calibration points. Figures 2 and 3 calibration images.

The K matrix is:

$$k = \begin{bmatrix} 2380.8 & 0 & 1277 \\ 0 & 2371.6 & 935.4 \\ 0 & 0 & 1 \end{bmatrix}$$

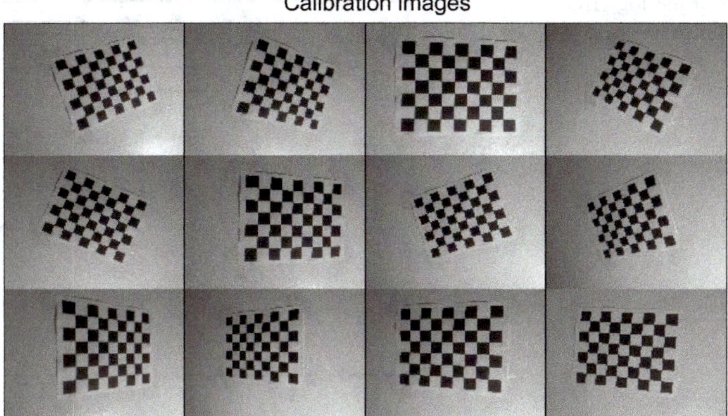

Fig. 2 Camera calibration toolbox for MATLAB with a model chessboard Plane

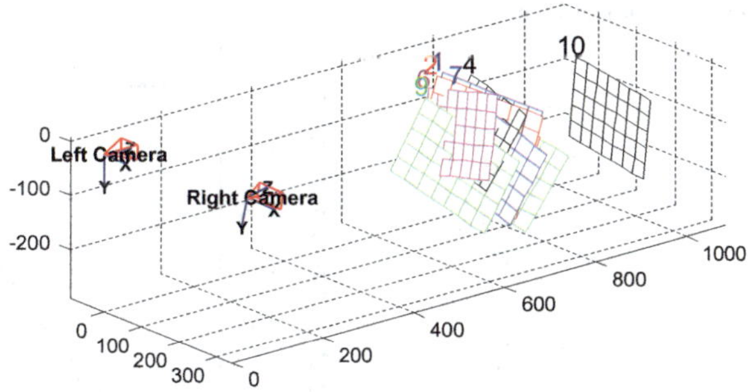

Fig. 3 Camera extrinsic parameters

6.2 The Error Function

To evaluate the algorithms, one can use the error in the second image after transferring points from the first image using the computed homography matrix. All the methods surveyed have been programmed and tested under the same conditions of image noise. The Euclidean image is the distance in the second image between the measured point x_i^* and the point H_{xi}. If d is the Euclidean distance between the inhomogeneous points, the transfer error in the set of correspondences is

$$E = \sum_i d(x_i^*, Hx_i)^2 \tag{6.1}$$

The performance of the three algorithms (DLT, RANSAC and Pseudo-inverse) in the presence of four level Gaussian noise was compared in Fig. 4 all the algorithms degrade as the noise level increases. RANSAC algorithm gives the best results among all the algorithms used.

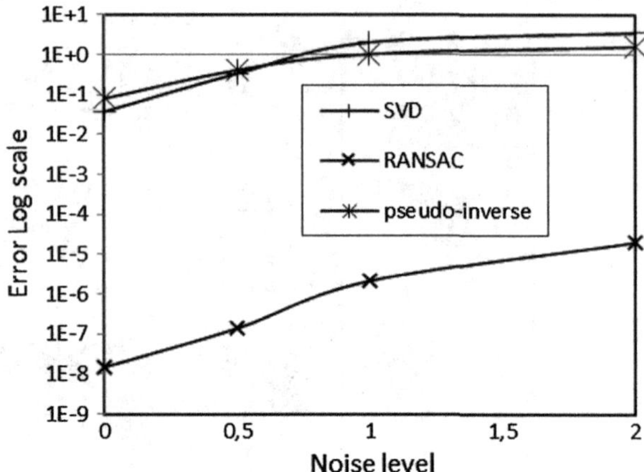

Fig. 4 Plot of the log error against the noise levels for all algorithms

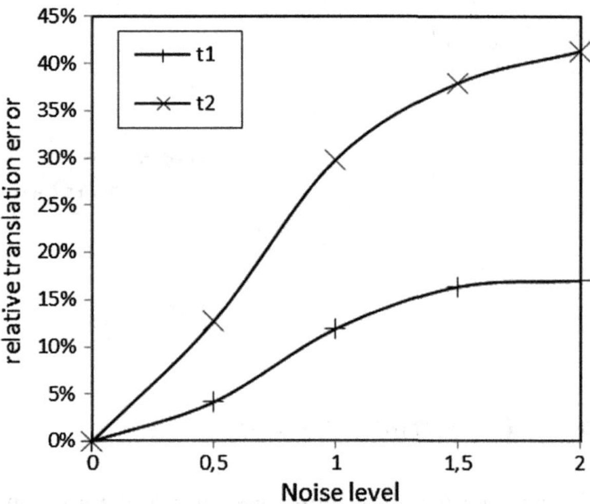

Fig. 5 Relative translation error via noise level

To verify the performance of the decomposition homography algorithm, we add the same noise error to the second image and calculated the relative translation error as shown in Fig. 5 it is a clear that all the error increase with the level of noise added to the image feature points.

6.3 *The Pose Measurements*

In this work, the accuracy and robustness of a pose estimate were performed using 2D homography, ICP algorithm, and Horn's absolute orientation method. All data were taken from two images in both front and top view as shown in Figs. 6 and 7

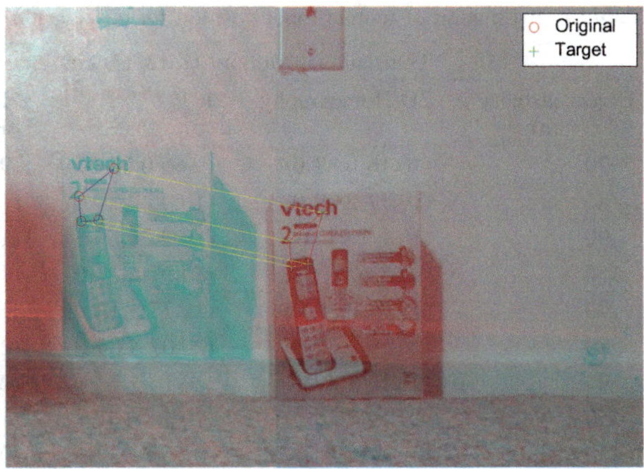

Fig. 6 Initial and desired target from front view

Fig. 7 Initial and desired target from top view

respectively, then it was compared with respect to the relative motion parameters. These algorithms were studied in Tables 1, 2 and 3 with varying distances and angles between the cameras. The data which gathering from the mobile robot was presenting in Tables 1 and 2 while the data which collect by flying robot are represented in Table 3. The distortion was increased as the distance and angle were increased. We noticed from Tables 1 and 3 that at the actual distance (10 0 0) the t1 is not affected while t2 increases when object distance increases. When the actual distance become (20 0 0) the t1 is not affected while t2 decreases when object distance increases. When the baseline or distance increased, both t1 and t2 are affected as shown in Table 1.

Table 1 Estimation of mobile robot position for all algorithms

	Translational Motion (t1, t2, t3) cm			
Object distance for z (cm)	2D Homography	ICP	Horn's absolute orientation method	Actual
70	(10.16 0.88 0)	(9.66 0.62 0)	(9.99 0.26 0)	(10 0 0)
80	(10.17 2.39 0)	(10.06 2.27 0)	(10.02 2.50 0)	(10 0 0)
90	(10.1 2.98 0)	(10 3.26 0)	(10.07 3.07 0)	(10 0 0)
150	(9.63 6.76 0)	(10.07 6.77 0)	(10.03 6.82 0)	(10 0 0)
70	(19.79 3.39 0)	(19.77 5.12 0)	(19.83 3.97 0)	(20 0 0)
80	(20.29 3.47 0)	(20.07 4.05 0)	(20.29 3.43 0)	(20 0 0)
90	(20.18 1.4 0)	(20.16 1.22 0)	(20.08 1.54 0)	(20 0 0)
150	(19.19 1.18 0)	(20.4 1.16 0)	(19.47 1.06 0)	(20 0 0)
150	(0.17 5.22 0)	(0.76 4.23 0)	(0.12 5.4 0)	(0 5 0)
150	(22.6 6 0)	(22.95 5.5 0)	(22.8 6.2 0)	(20 10 0)
150	(19.7 3.58 0)	(19.69 2.65 0)	(19.7 3.05 0)	(20 5 0)
300	(28.4 2.1 0)	(28.62 1.5 0)	(27.4 3.05 0)	(30 0 0)
500	(37.8 3.3 0)	(38.1 3.1 0)	(36.6 4.2 0)	(40 0 0)
1000	(45.5 4.3 0)	(46.3 4.01 0)	(44.7 4.5 0)	(50 0 0)

Table 2 Estimation of mobile robot position for two baseline

Object distance for Z(cm)	Baseline = 10 cm			Baseline = 20 cm		
	2D Homography	ICP	Horn's absolute orientation method	2D Homography	ICP	Horn's absolute orientation method
70	(10.16 0.88 0)	(9.66 0.62 0)	(9.99 0.26 0)	(19.79 3.39 0)	(19.77 5.12 0)	(19.83 3.97 0)
80	(10.17 2.39 0)	(10.06 2.27 0)	(10.02 2.50 0)	(20.29 3.47 0)	(20.07 4.05 0)	(20.29 3.43 0)
90	(10.1 2.98 0)	(10 3.26 0)	(10.07 3.07 0)	(20.18 1.4 0)	(20.16 1.22 0)	(20.08 1.54 0)
150	(9.63 6.76 0)	(10.07 6.77 0)	(10.03 6.82 0)	(19.19 1.18 0)	(20.4 1.16 0)	(19.47 1.06 0)

Table 3 Estimation of flying robot position for two baseline

Object ditance for Z(cm)	Baseline = 10 cm			Baseline = 20 cm		
	2D Homography	ICP	Horn's absolute orientation method	2D Homography	ICP	Horn's absolute orientation method
60	(10.22 0.33 0)	(9.8 0.32 0)	(10.1 0.56 0)	(19.6 2.11 0)	(19.54 4.32 0)	(19.98 3.2 0)
80	(10.21 3.13 0)	(10.1 2.15 0)	(10.12 3.1 0)	(20.41 4.13 0)	(20.18 4.37 0)	(20.40 4. 3 0)
120	(10.3 3.19 0)	(10.2 3.7 0)	(10.11 3.26 0)	(20.38 3.5 0)	(20.3 2.41 0)	(20.21 2.66 0)
150	(9.32 7.16 0)	(10.21 5.87 0)	(10.1 7.14 0)	(19.43 2.8 0)	(20.5 2.7 0)	(19.51 2. 6 0)

7 Conclusions

The objective of this paper is a motion estimation and accurate pose estimation of mobile and flying robots using a stereo camera. Vision based navigation is applied for the mobile robot to navigate and avoid the obstacle in indoor environment, to flying robot aerial refueling, and to space robot for debris clearance operations. Many different techniques are used to estimate the planar homography between two images. The homography matrix needs only four features are laid on the different plane while the fundamental matrix needs more than eight features, which are laid on the same plane. The literature on homography estimation is quite extensive, and many advanced techniques are available. The Random Sample Consensus (RANSAC) gives the stability of the results with respect to the noise levels while the other methods were strongly influenced by the noise levels. Furthermore, the smallest relative translation error has occurred when the noise level is less than 0.5. This error was approximately stable when the noise error is below 0.5.

Three algorithms (2D Homography, ICP, and Horn's Absolute Orientation method) were studied to find the accuracy and robustness of each algorithm and to find the optimal baseline distance between lenses. The distortion increases as the z-distance or baseline are changed. These errors come from a distortion in images or from inaccurate camera calibration. It could be seen in Tables 2 and 3 that accurately a baseline is 10 cm when one needs to measure the object distance equal or less than 80 cm while the most accurate baseline is 20 cm when there is a need to measure the distance more than 80 cm. From Table 1, it can be noticed that the error was increased when the baseline or object distance increased.

References

Besl P, McKay N (1992) A method for registration of 3-D shapes. IEEE Trans Pattern Anal Mach Intell 14(2):239–256

Bouguet J (2015) Camera calibration toolbox for matlab. www.vision.caltech.edu/bouguetj/calib_doc

DeMenthon DF, Davis LS (1992) Model-based object pose. In: 25 lines code. European conference on computer vision, pp. 335–343

Faugeras O, Lustman F (1988) Motion and structure from motion in a piecewise planar environment. Int J Pattern Recogn Artif Intell 485–508

Horn BKP (1987) Closed-form solution of absolute orientation using unit quaternions. JOSA A 4 (4):629–642

Hu G, MacKunis W, Gans N, Dixon W, Chen J, Behal A, Dawson D (2009) Homography-based visual servo control with imperfect camera calibration. IEEE Trans Autom Control 54 (6):1318–1324

Kniaz VV (2016) Robust vision-based pose estimation algorithm for an UAV with known gravity vector. Int Arch Photogrammetry Remote Sens Spat Inf Sci XLI-B5:63–68

Ma Y, Soatto S, Kosecka J, Sastry SS (2003) An invitation to 3-D vision: from images to geometric models. Springer, Berlin

Michaelson E, Kirchhof M, Stilla U (2004) Sensor pose inference from airborne videos by decomposing homography estimates. In: Proceedings of the XXth ISPRS congress, technical commission III, Istanbul, Turkey

Montijano E, Sagues C (2009) Fast pose estimation for visual navigation using homographies. In: IEEE/RSJ international conference on intelligent robots and systems, October 11–15, pp 356–361

Quan L, Lan Z (1999) Linear n-point camera pose determination. IEEE Trans Pattern Anal Mach Intell 21(8):774–780

Walker M, Sasiadek JZ (2013) Accurate pose determination for autonomous vehicle navigation. In: IEEE/conference on methods and models in automation and robotics, pp 356–361

Walker M, Sasiadek JZ (2015) Accurate image depth determination for autonomous vehicle navigation. In: CARO3—3rd conference on aerospace robotics

Xu Y, Luo D, Xian N, Duan H (2014) Pose estimation for UAV aerial refueling with serious turbulences based on extended Kalman filter. Optik Int J Light Electron Optics 125 (13):3102–3106

Xuebo Z (2008) A fast homography decomposition technique for visual servo of mobile robots. In: Proceedings of the 27th Chinese control conference, Kunming, Yunnan, China, July 16–18

Control of Flexible Wing UAV Using Stereo Camera

Malik M. A. Al-Isawi and Jurek Z. Sasiadek

1 Introduction

Ultralight gliders, also known as flex-wing trikes, microlight trikes, or microlights, have attracted immense attention in the air sport community since 1980s. In recent years, commercial companies and military bases have shown their interests to utilize such vehicles since they are relatively more affordable and have lower maintenance and performance costs. These aerial vehicles can carry up to two passengers (including the pilot) and are equipped with simple flight instruments for navigational purposes. The flexibility of such an aircraft leads to material deformation and linear theories are not relevant for their analysis. In fact, the effect of the deformations of the aircraft positively affects the flight stability and control characteristics. There has been very limited research around classification of wing deflection and around control of a flexible wing by using a vision system. Some of the works were studied on aerodynamics coefficients and aeroelasticity of flying. The authors Weisshaar and Ashley (1973) were focused on static aeroelasticity of flying wings. In Patil and Hodges (2006) a theoretical study for the dynamic response estimation of a highly flexible flying-wing is described. Aerodynamic coefficients and structural displacements from wind tunnel tests are presented Albertani et al. (2005) also the wing flexibility on a MAV are investigated. The videogrammetric technique (Burnerand and Liu 2001) have been presented and measured the deformation and attitude on static and dynamic in a wind-tunnel model. The non-contact videogrammetry technique was measured and studied the dynamic behavior of flexible mini-UAV wings in wind tunnel tests (Pitcher et al.

M. M. A. Al-Isawi (✉)
Al-Khwarizmi Engineering College, University of Baghdad, Baghdad, Iraq
e-mail: malikalisawi@cmail.carleton.ca

J. Z. Sasiadek
Department of Mechanical and Aerospace Engineering, Carleton University, Ottawa, Canada
e-mail: JurekSasiadek@cmail.carleton.ca

J. Sasiadek, *Aerospace Robotics III*, GeoPlanet: Earth and Planetary Sciences,
https://doi.org/10.1007/978-3-319-94517-0_7

2009). The authors Burner et al. (1996) presented the wing twist technique based upon a camera photogrammetric. The wing structure deformation was determined by Bakunowicz and Meyer (2016). They were collecting the measurement data in flight test by using the digital image correlation. The static aeroelasticity was measured and discussed by Burner et al. (2002). In addition, the uncertainty of videogrammetric technique was used in the wind tunnel model. A fuzzy logic implemented by Chiu et al. (2002) was used to determine the surface deflections for the flexible wing to achieve the desired roll angle. The authors Kurnaz et al. (2010) described the ANFIS algorithm for UAV to enable it to autonomously accomplish its mission. The lateral controller is developed by Farid and Barakati (2013). They used ANFIS and compared it with PID controller for UAVs latitude.

The goal of this paper is the development and implementation of a new approach that would extend the information gathered by a stereo camera from flight tests or wind tunnel for the flexible wing to design a system that has the ability to guide and control of UAV. This system was achieved the relationship between wing flexibility and flight performance of the wing such as straight, level flight, and turns. The platform selected for this research is the wing of the hang glider. This platform was chosen for its relatively more affordable and easy maintenance and performance costs. A fuzzy logic was used to classify the wing shape depending on the deformation of a flexible wing. In addition, an ANFIS controller was used to achieve the desired flight performance.

2 Stereo Vision Camera

The stereo vision is the process of extracting the depth of a 3-D scene from different points of view. The stereo camera has two cameras with a horizontally aligned and fixed distance known as the baseline. The ZED, stereo camera was chosen as shown in Fig. (1) for providing vision data as well as depth data. One of the reasons to choose this camera is that it can capture a high-resolution side by side video on USB3. Furthermore, this camera is already calibrated and comes with known intrinsic and extrinsic parameters. The epipolar constraints are using to reduce the time that is needed to search for corresponding points in two images. This can result in searching just a line of the image instead of a full size image (Loop and Zhang 1999).

Fig. 1 StereoLabs ZED camera

2.1 Modeling of Stereo Camera

Firstly, it is convenient to define the geometry projection model of the stereo camera. The homography can be computed from the relationship between the two locations using the 2D pixel information (feature matching point) and the camera parameters (camera calibration) as shown in Fig. (2). The feature matching process has two steps. The features are extracted from two cameras using SURF methods, then the distinguish features are matched.

The mathematical model of the stereo vision is as follows. Assuming a feature point, F has the corresponding pixel points the scene, also, F_l and F_r are the corresponding pixel points of F in the left and right image respectively, and it acts at the same height level depend on the epipolar constraint as shown in Fig. 3. The depth equation can be derived based on the geometric of similar triangles,

$$Z = \frac{fb}{(X_l - X_r)} = \frac{fb}{d} \tag{1}$$

where b is baseline, f is a focal length, and d is disparity. The baseline and focal length of our stereo camera are known, so, the distance Z in the scene can be calculated based on the disparity in the two images.

The corresponding points F1 and F2 are related by the homography matrix (Xuebo 2008; Al-Isawi and Sasiadek 2015)

$$F_1 \approx HF_2 \tag{2}$$

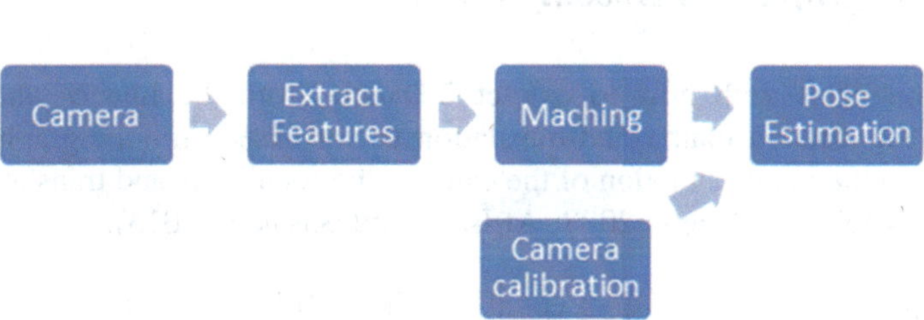

Fig. 2 Steps of pose estimation

Fig. 3 Mathematical of a stereo vision

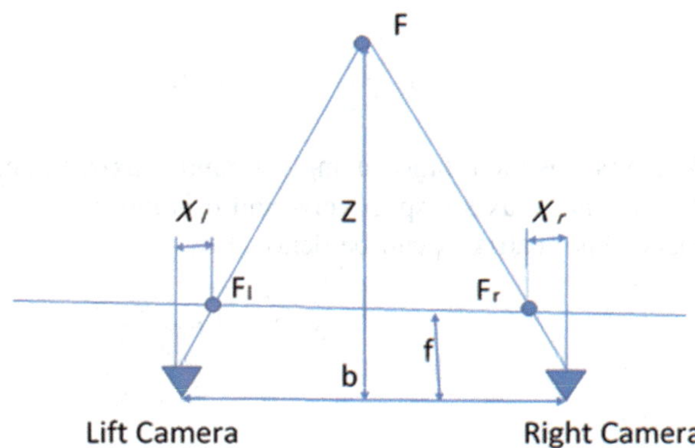

The equation of homography is Faugeras and Lustman (1988):

$$H = R - \frac{t}{d} n^T \tag{3}$$

where H is homography matrix and can be decomposed to R, t/d and n., the R is a rotation matrix, t is the translation vector of the camera, d is the distance between the camera to the plane, and n is the surface normal vector to the plane (Faugeras and Lustman 1988). The Euclidean coordinates for F_1 and F_2 can be expressed in image coordinates.

$$P_1 = kF_1 \text{ and } P_2 = kF_2 \tag{4}$$

where, $k \epsilon R^{3\times3}$ is a constant camera matrix and it was determined from camera calibration (Al-Isawi and Sasiadek 2015),

$$k = \begin{bmatrix} f_x & \beta & u_o \\ 0 & f_y & v_o \\ 0 & 0 & 1 \end{bmatrix} \tag{5}$$

where, k is an intrinsic camera matrix, (f_x, f_y) is the focal length in x and y directions respectively, u_0 and v_0 are the principal points in the pixel dimensions. β is the skew coefficient between the camera's *x*- and *y*-axes, which is often zero. F_1 and F_2 are pixel coordinates of two images.

2.2 *Homographics's Decomposition*

This section deals with how to extract 2-D data from the images and how to determine the rotation matrix and translation vector based on homography matrix. According to the planner motion of the camera, the rotation R and translation T can be shown as follows (Xuebo 2008; Al-Isawi and Sasiadek 2015):

$$R = \begin{bmatrix} \cos\theta & -\sin\theta & 0 \\ \sin\theta & \cos\theta & 0 \\ 0 & 0 & 1 \end{bmatrix} \tag{6}$$

$$T = \frac{t}{d} = \begin{bmatrix} \frac{t_x}{d} & \frac{t_y}{d} & 0 \end{bmatrix}^T = \begin{bmatrix} T_1 & T_2 & 0 \end{bmatrix}^T \tag{7}$$

where $\theta \epsilon R$ is the rotational angle around Z-axis, t_x, and t_y are the translations along X-axis and Y-axis respectively, and d is the distance from origin to the reference plane. The matrix H can be denoted by:

$$H = \begin{bmatrix} \cos(\theta) + T_1 n_1 & -\sin(\theta) + T_1 n_2 & T_1 n_3 \\ \sin(\theta) + T_2 n_1 & \cos(\theta) + T_2 n_2 & T_2 n_3 \\ 0 & 0 & 1 \end{bmatrix} = \begin{bmatrix} h_{11} & h_{12} & h_{13} \\ h_{21} & h_{22} & h_{23} \\ 0 & 0 & 1 \end{bmatrix} \quad (8)$$

It's clear from (8), that the algorithm for homography decomposition will be divided into three cases depending on the h13 and h23 (Faugeras and Lustman 1988).

$$T_1 = \varepsilon_1 \sqrt{\frac{h_{13}^2[(h_{12} + h_{21})^2 + (h_{11} - h_{22})^2}{h_{13}^2 + h_{23}^2} + h_{13}^2} \quad (9)$$

$$T_2 = \varepsilon_1 sgn(h_{23} h_{13}) \sqrt{\frac{h_{23}^2[(h_{12} + h_{21})^2 + (h_{11} - h_{22})^2}{h_{13}^2 + h_{23}^2} + h_{23}^2} \quad (10)$$

where $\varepsilon_1 = \pm 1$, it can be noted that there are two solutions of T that can be chosen according to Farid and Barakati (2013). Also, we can compute the n vector in (8) as follows:

$$n_1 = \frac{t_1(h_{11} - h_{22}) + t_2(h_{12} + h_{21})}{t_1^2 + t_2^2} \quad (11)$$

$$n_2 = \frac{t_1(h_{12} + h_{21}) - t_2(h_{11} - h_{22})}{t_1^2 + t_2^2} \quad (12)$$

$$n_3 = \varepsilon_1 sgn(h_{13}) \sqrt{\frac{h_{13}^2 + h_{23}^2}{t_1^2 + t_2^2}} \quad (13)$$

It's easy to compute the rotation matrix from (3).

3 Classification Using the Fuzzy Logic

The fuzzy logic approach has many different models to classify objects. In general, the fuzzy logic approach has three steps: fuzzification, fuzzy inference, and defuzzification. The fuzzification step converts input value to a linguistic value. The fuzzy inference is responsible of formulating the mapping from a crisp input value to linguistic value using fuzzy rules. This process involves membership function, and the operation of fuzzy logic. The defuzzification process converts the fuzzy output into a crisp value (Nedeljkovic 2004).

4 Control System Design

The proposed controller is implemented in form of two different controllers, namely LQR controller, and ANFIS controller. These controllers were designed to ensure that the platform will be stable within a certain threshold. The airspeed, AOA, and roll angle are controlled by using the combination of two inputs, control bar, and throttle angle.

4.1 LQR Controller

LQR controller is an optimal control approach based on feedback control (Rahimi et al. 2013). In LQR, a cost function for optimal control performance is:

$$J = \int_{0}^{\infty} \left(x^T Qx + u^T Ru\right) dt \tag{14}$$

It is important to know that the vector u in Eq. (14) minimizes the quadratic cost function which leads to optimal feedback control law represented as (Rahimi et al. 2013).

$$\mathrm{u} = -\mathrm{Kx} \tag{15}$$

$$K = R^{-1} BP \tag{16}$$

The steady state optimal gain K is determined by using the Riccati equation as below

$$A^T P + PA - PBR^{-1} B^T P + Q = 0 \tag{17}$$

where, Q and R are positive semi-definite and positive definite, respectively. Matrix Q is known as a cost matrix and R is performance index matrix. The weight matrices of Q and R are very important in LQR method, and they are symmetric and nonnegative matrices. The control performance is affected by the weight matrices. These matrices are determined by experience of engineers who are familiar with the controlled system (Tamaskani et al. 2015).

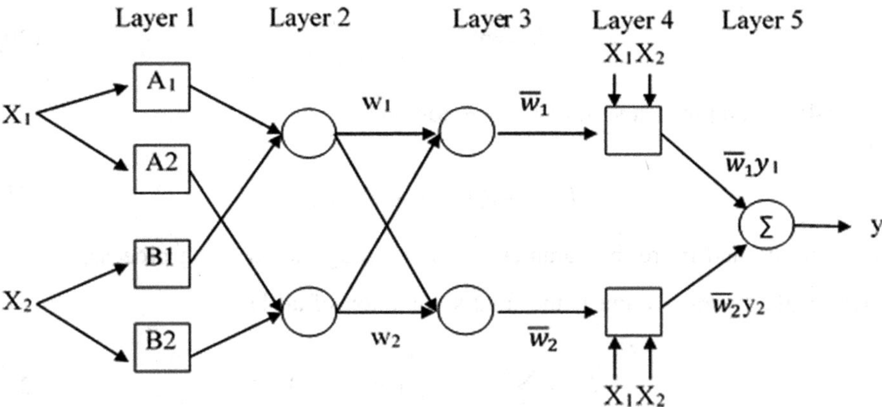

Fig. 4 Adaptive neuro fuzzy interference system structure

4.2 *ANFIS Controller*

A Neuro-Fuzzy system has been proposed by Jang (1993), Mitra and Hayashi (2000). Adaptive Neuro-Fuzzy Inference Systems (ANFIS systems) is containing the fuzzy logic and the neural network. The hybrid learning algorithm is used for training the membership function and the fuzzy rules at any number or shape of membership functions with less training times.

The ANFIS structure has five layers as shown in Fig. 4.

Layer 1: the output for this layer (L_i^1) is a membership function (μ_{Ai}) of each input variable (xi)

$$L_i^1 = \mu_{Ai}(x_1) \tag{18}$$

Layer 2: It is the product of the two inputs

$$L_i^2 = w_i = \mu_{Ai}(x_1) \cdot \mu_{Bi}(x_2) \tag{19}$$

Layer 3: Each node in this layer determine the normalized firing strength as follow:

$$L_i^3 = \bar{w}_i = \frac{w_i}{w_1 + w_2} \tag{20}$$

Layer 4: the output is a sum of the previous ones such as:

$$L_i^4 = \bar{w}_i(p_i x_1 + q_i x_2 + r_i) \tag{21}$$

where pi, qi, and ri are the parameters come through the training process.

Layer 5: the overall output is the final summation of all layers.

$$L_i^5 = \sum_i \bar{w}_i(p_i x_1 + q_i x_2 + r_i) \tag{22}$$

5 Results

The novel contribution of this paper starts with capturing images and extracting the distinguished feature of flexible wing. Subsequently, 3D points were computed for each particular point with the help of vision system, as described in the previous section. After that, we used a Fuzzy set associated with each point to classify the shape. Furthermore, ANFIS and LQR controllers were used to control the flying vehicle.

5.1 Extract Deflection by Using Stereo Camera

Experimental testing was performed in this paper to clarify the proposed approach of deflection measurement of the flexible wing using a stereo camera. UAV platform which was used in wind tunnel testing has flexible wing 120 cm wingspan and 100 cm long also it has no internal components like fuselage and engine for the propeller. The flexible wing is composed of a plastic fiber while the structure body is made from wood. The advantages of using flexible wing are that it has greater resistance at high AOA. Besides that, roll stability increased. Six square markers attached at the end of the wing as shown in Figs. (5, 6 and 7) are used to gather the measurements data from a stereo camera.

Fig. 5 Front view of flexible wing in wind tunnel

Fig. 6 Back view of flexible wing in wind tunnel

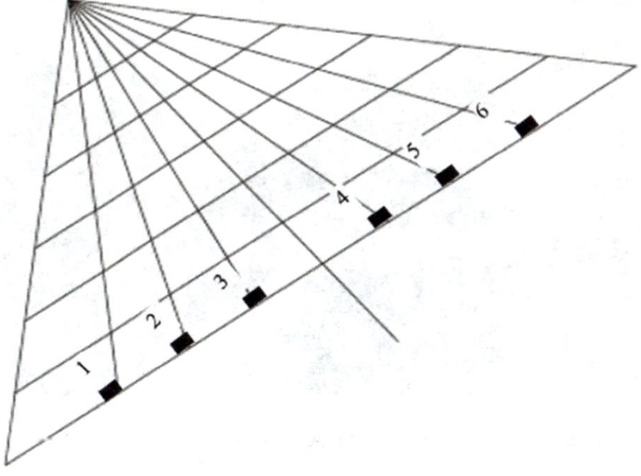

Fig. 7 Feature's position

It is fixed behind the wing on the keel tube to ensure that the wing is in the view. The ZED, stereo camera is used for providing vision data for square markers. The experimental test was beginning to measure the deflection of wing. Figures 5 and 6 show the back and front view of the wing. The test procedures were first taken pictures of the wing at zero AOA and zero roll angle, with specific wind velocity (11, 21, 31) Km/h as shown in Fig. (8). After that the value of AOA for wing was changed to +20°, the roll angle was unchanged, and the camera took pictures for the same values of wind velocity as shown in Figs. (9, 10 and 11). The test was repeated for AOA = −20° at the same conditions as shown in Fig. (12). Moreover, the roll angle was changed to 10° and AOA to 20° with the same selected wind velocities as shown in Fig. (13). Finally, the roll angle was set to −10° and AOA to −20° and pictures were taken for different wind values velocity as shown in Fig. (14).

Fig. 8 Shape of wing at V = 31 km/h

Fig. 9 Shape of wing at V = 11 km/h and AOA = +20°

Fig. 10 Shape of wing at V = 21 km/h and AOA = +20°

Fig. 11 Shape of wing at V = 31 km/h and AOA = +20°

Fig. 12 Shape of wing at V = 31 km/h and AOA = −20°

Fig. 13 Shape of wing at V = 31 km/h, Roll angle = 10° and AOA = 20°

Fig. 14 Shape of wing at V = 31 km/h, Roll angle = −10° and AOA = −20°

Several shapes of the flexible wing were observed throughout the wind tunnel. At AOA = 0 and roll angle = 0, the deflection increased to −0.5 cm when the wind velocity reached to 31 km/h, and it was noticed that the shape of the wing was symmetry, and the peak deflection occurred at the mid distance of each side of the wing as shown in Fig. (15). When AOA became −20°, the deflection decreased to −0.15, and we noticed that if the shape of the wing was symmetry on both sides, then the roll angle is zero, or if it is not, then the roll angle is not zero as shown in Figs. (16) and (17). From observation, the maximum deflection occurred when AOA was equal to 20°. Figure (18) shown that the roll angle is zero because the shape of the wing was the similarity in both sides while Fig. (19) shows the shape of the wing is not similar that means the roll angle was not zero, and the value of this angle depends on the value of deflection.

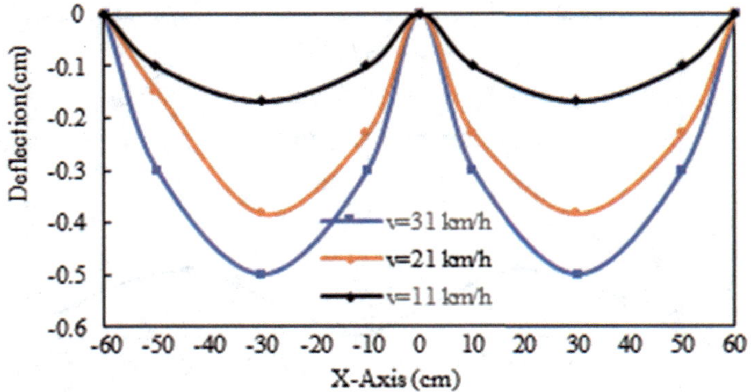

Fig. 15 Shape of wing at AOA = 0° and roll angle = 0°

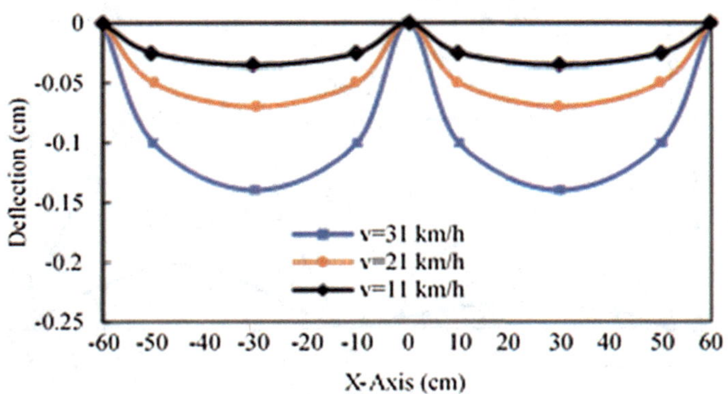

Fig. 16 Shape of wing at AOA = −20° and roll angle = 0°

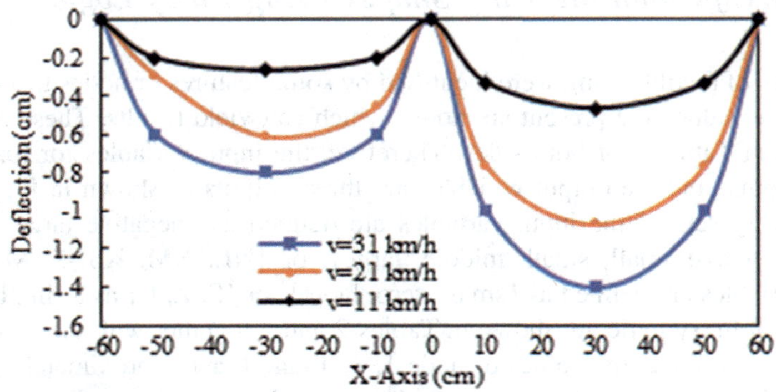

Fig. 17 Shape of wing at AOA = −20° and roll angle = −10°

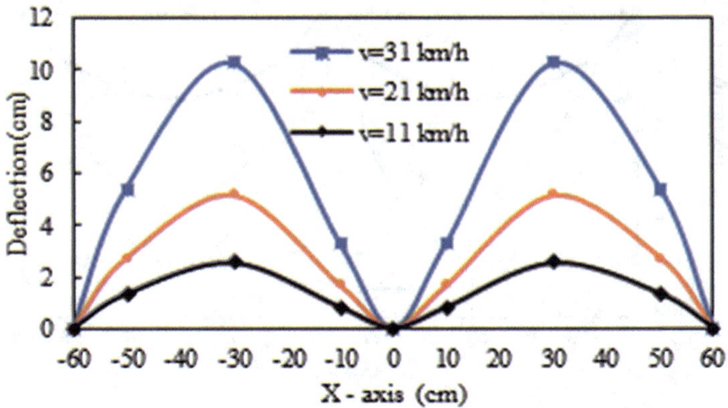

Fig. 18 Shape of wing at AOA = 20° and roll angle = 0°

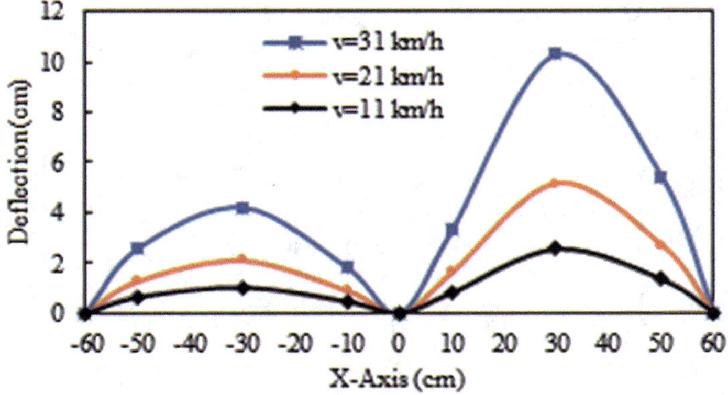

Fig. 19 Shape of wing at AOA = 20° and roll angle = 10°

5.2 Classification the Wing Shapes Using Fuzzy Logic

The shapes of flexible wing were identified by some features or nodes that each one of them has value. We present six nodes which can yield results. These nodes are presented in Table 1 for both sides. Therefore, the input variables for fuzzy logic are six inputs, and the output variables are three outputs as shown in Fig. 20.

The fuzzy sets of the input variables are defined as {negative large, negative middle, negative small, small, middle, large}, or {NL, NM, NS, S, M, L} and output variables are defined as {small, zero, large}, or {S, Z, L} as a simplification. The rules of fuzzy logic are shown in Tables 2 and 3 together with the membership function used. Since, the values of node 1, 3, 4, and 6 are approximately the same and node 2 and node 5 have the same values described in Table 1. The node 1, and

Table 1 Nodes deflection over the both sides

Classical input		Deflection in cm					
	Node No.	NS	NM	NL	S	M	L
Left side of wing	1	−0.025	−0.23	−1	0.65	2.6	5.4
	2	−0.035	−0.385	−1.4	1.05	4.2	10.3
	3	−0.025	−0.23	−1	0.45	1.8	3.3
Right side of wing	4	−0.025	−0.23	−1	0.45	1.8	3.3
	5	−0.035	−0.385	−1.4	1.05	4.2	10.3
	6	−0.025	−0.23	−1	0.65	2.6	5.4

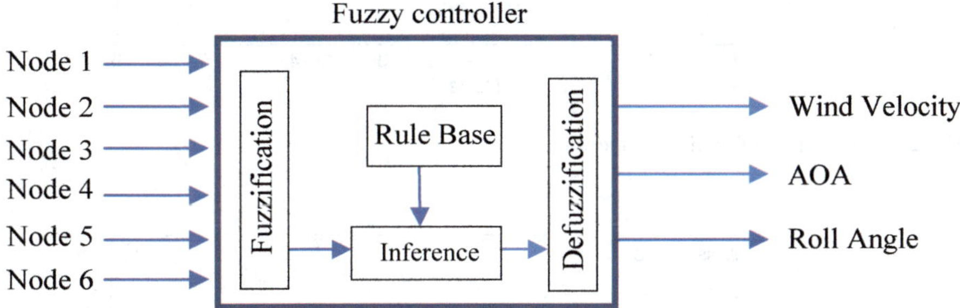

Fig. 20 Block diagram of fuzzy logic

Table 2 Fuzzy rules for the flexible wing at a positive deflection

V, AOA, R		Node 1, 2, 3		
		S	M	L
Node 4, 5, 6	S	S, M, Z	M, P, N	L, P, N
	M	M, P, P	M, M, Z	M, P, N
	L	L, P, P	M, P, P	L, L, Z

Table 3 Fuzzy rules for the flexible wing at a negative deflection

V, AOA, R		Node 1, 2, 3		
		NL	*NM*	*NS*
Node 4, 5, 6	NL	L, Z, Z	M, N, N	L, N, N
	NM	M, N, P	M, M, Z	M, Z, N
	NS	L, N, P	M, Z, P	S, N, Z

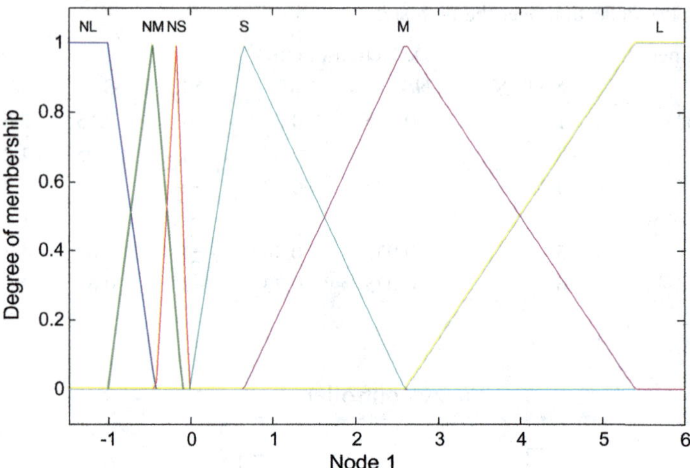

Fig. 21 Membership function of the input node 1

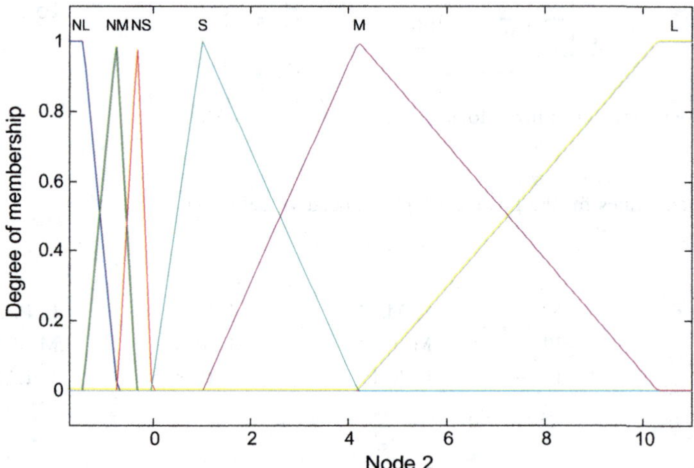

Fig. 22 Membership function of the input node 2

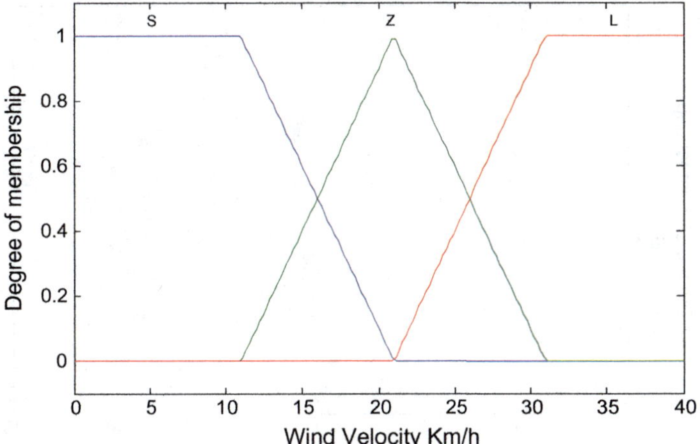

Fig. 23 Membership function of the output wind speed

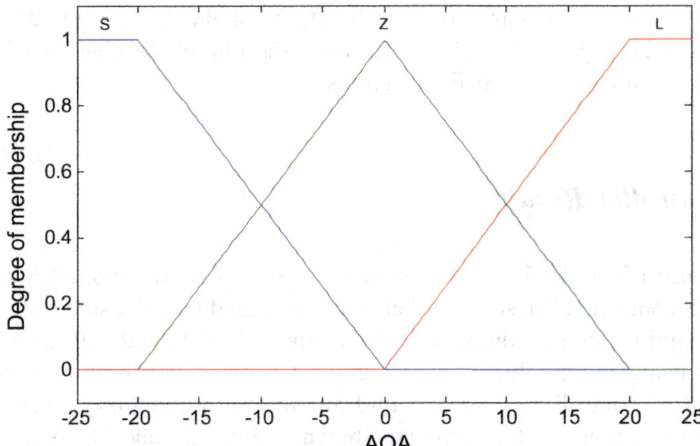

Fig. 24 Membership function of the output AOA

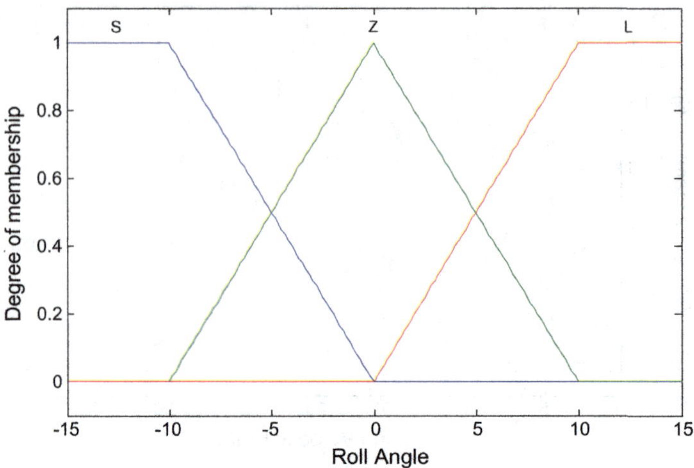

Fig. 25 Membership function of the output Roll angle

node 2 were chosen as a membership function for the fuzzy logic as shown in Figs. 21 and 22. Figures 23, 24 and 25 show the output variables wind velocity, AOA, and roll angle membership functions.

5.3 Controller Results

The proposed controller is implemented through two controllers, LQR controller, and ANFIS controller. These controllers were designed to make sure stabilize of the platform within a certain threshold. The airspeed, AOA, and roll angle are controlled by using the combination of a control bar and the throttle angle. In this paper, ANFIS controller is used to find automatically the fuzzy rules shown in Fig. 26 and membership functions are shown in Figs. 27 and 28. A Sugeno type

Fig. 26 3D surface for ANFIS rules

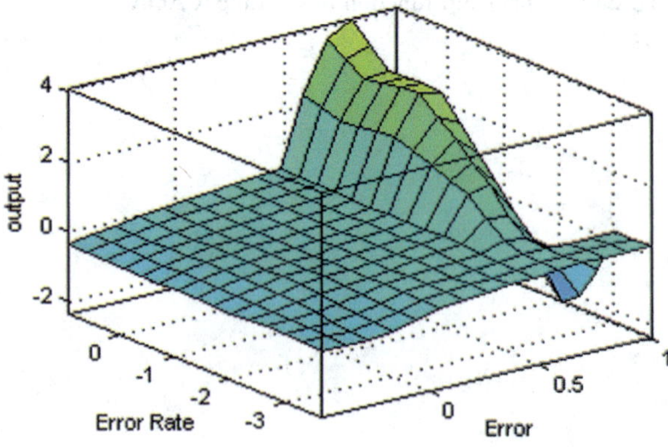

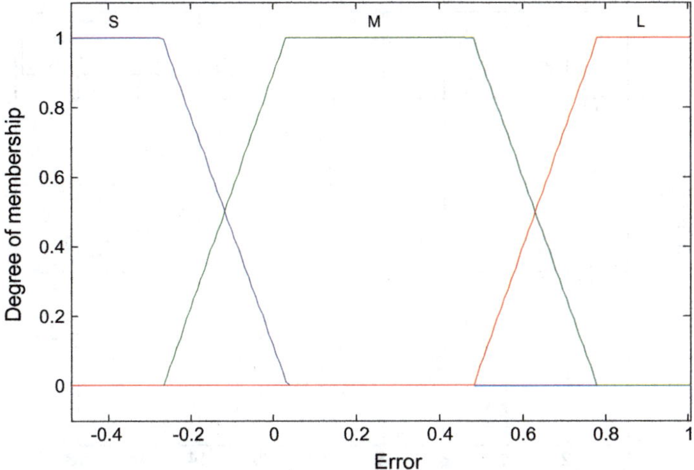

Fig. 27 Membership function of the error in node 1

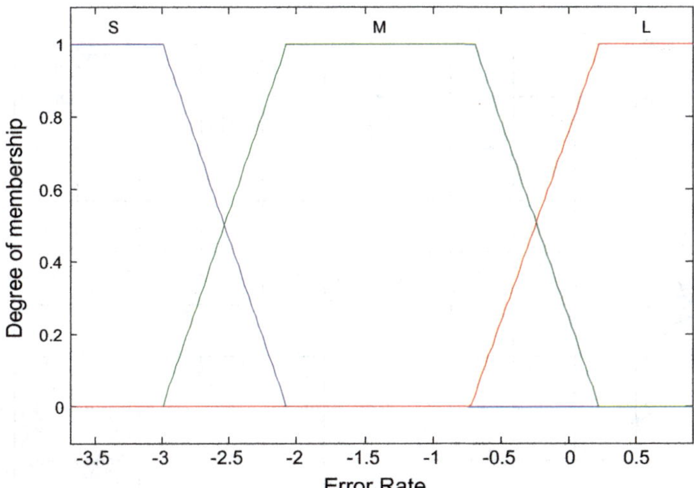

Fig. 28 Membership function of the error rate in node 1

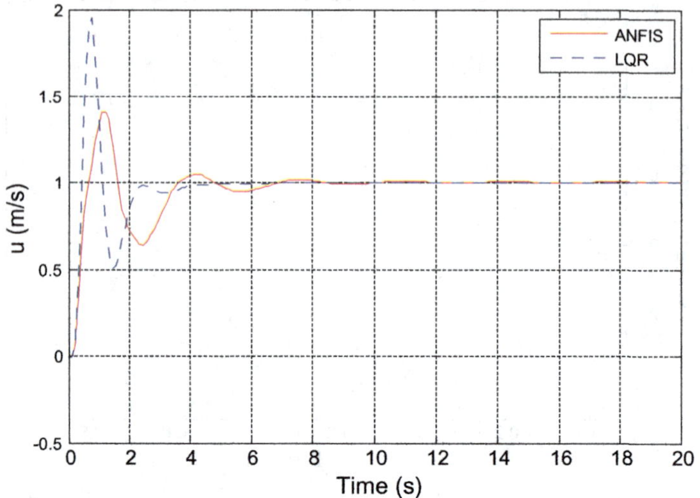

Fig. 29 Airspeed with step change in control bar input

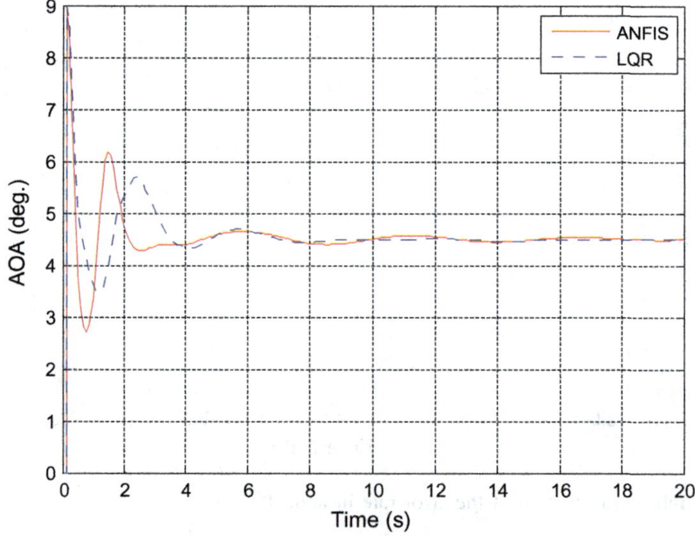

Fig. 30 AOA with step change in control bar input

fuzzy inference system with two trapezoid membership function for inputs and one for output was used in ANFIS. LQR controller was also used, and the performance of both controllers has been investigated when applying step function as an input. Figure 29 shows the airspeed response when control bar is applied as an input. The rise time and settling time in LQR controller is better than ANFIS controller

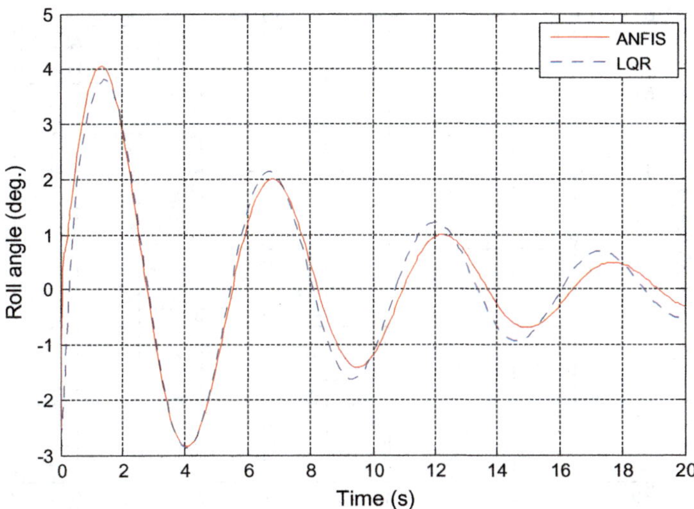

Fig. 31 Roll angle with step change in control bar input

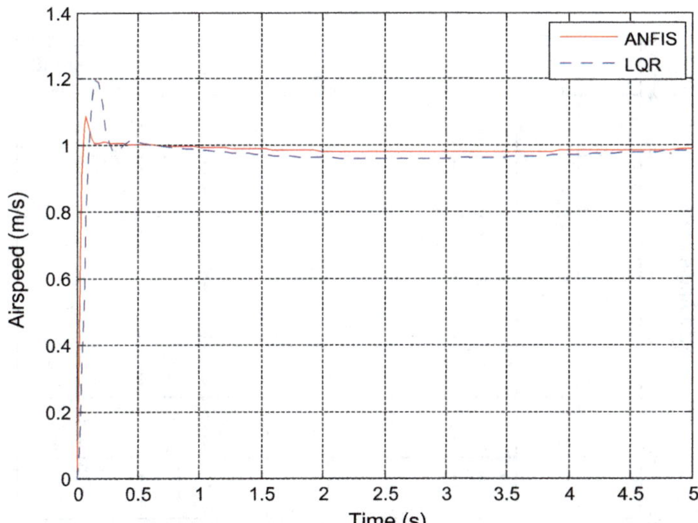

Fig. 32 Airspeed with step throttle input

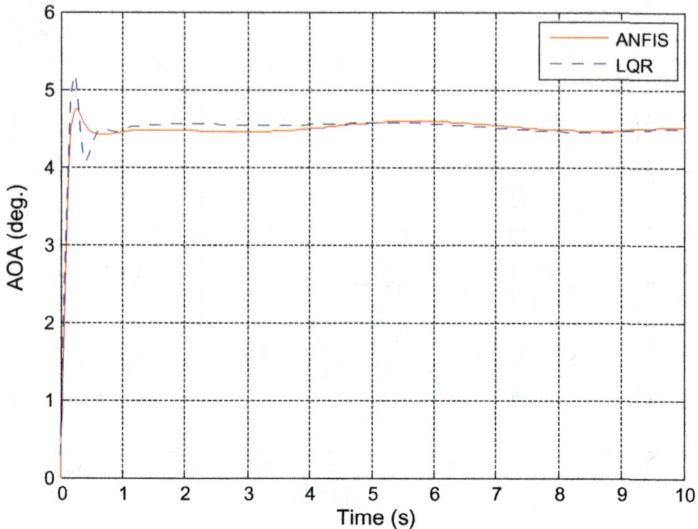

Fig. 33 AOA with step throttle input

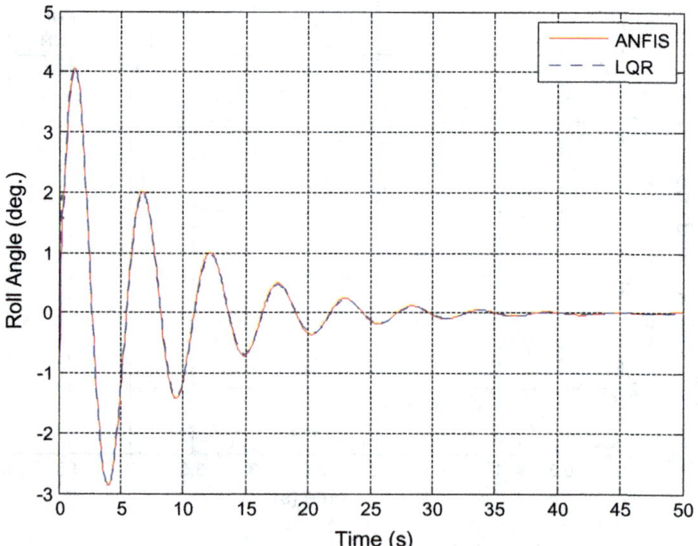

Fig. 34 Roll Angle with step throttle input

while the maximum overshoot in ANFIS is half than in LQR. The AOA and Roll angle are presented in Figs. 30 and 31 respectively where the platform is controlled by control bar. These figures show that both controllers are approximately the same response. ANFIS controller has a better response than the LQR controller for airspeed and AOA when we are using a throttle angle as input as shown in Figs. 32 and 33 while the roll angle in Fig. 34 has unchanged performance in both controllers.

6 Conclusion

The objective of this paper was to develop a control system for flexible wing Unmanned Aerial Vehicle (UAV). The presented method includes, identifying the deflection of the flexible wing and classifying its shape to find the dynamics characteristic for UAV. The proposed system is composed of three main elements; one of them is a vision system; the second is the fuzzy logic algorithm, and the third is Adaptive Nero-Fuzzy (ANFIS) controller. The vision system is using a ZED stereo camera in order to determine the shape of a flexible wing. The fuzzy logic system has been designed to classify the shape of the flexible wing and find the dynamic behavior for this wing. Several experiments were performed to verify the accuracy of this technique over a board range of wind speed, angle of attack (AOA), and Roll angle. The results from the wind tunnel show that the maximum wing deflection occurred when the wind speed is 31 km/h and AOA equal or greater than 0°, as shown in Figs. 18 and 19. The presented result demonstrates that vision system is an important part of an overall control system. The fuzzy logic system was using 18 rules for recognizing a suitable shape of the wing and to evaluate the system performance within a short period of time. The adaptive neuro-fuzzy controller (ANFIS) results were compared with the Linear Quadratic Controller (LQR) controller results. Both controllers did show a stable performance in airspeed, angle of attack (AOA), and roll angle. The input was introduced by control bar or throttle angle. The airspeed performance with LQR controller is faster when the control bar was used as input, but the overshoot error was greater than for ANFIS controller. The results also have shown that the ANFIS controller performed significantly better than LQR controller when the input was the throttle angle.

References

Albertani R, Stanford B, Hubner JP, Lind R, Ifju P (2005) Experimental analysis of deformation for flexible-wing micro air vehicles. In: 46th AIAA/ASME/ASCE/AHS/ASC structures, Structural dynamics & materials conference 18–21 Apr 2005

Al-Isawi MMA, Sasiadek JZ (2015) Pose estimation for mobile robot via visual navigation. In: Proceedings of the CARO3—3rd conference on aerospace robotics, Nov 2015

Bakunowicz J, Meyer R (2016) In-flight wing deformation measurements on a glider. Aeronaut J 120(1234):1917–1931

Burner AW, Wahls RA, Goad WK (1996) Wing twist measurements at the national transonic facility. NASA Technical Memorandum 110229, Feb 1996

Burner AW, Liu T, DeLoach R (2002) Uncertainty of videogrammetric techniques used for aerodynamic testing. In: 22nd AIAA aerodynamic measurement technology and ground testing conference, pp 24–26

Burnerand AW, Liu T (2001) Videogrammetric model deformation measurement technique. J Aircr 38(4):745–754

Chiu S, Chand S, Moore D, Chaudhary A (2002) Fuzzy logic for control of roll and moment for a flexible wing aircraft. IEEE Control Syst 42–48

Farid AM, Barakati SM (2013) UAV controller based on adaptive neuro-fuzzy inference system and PID. Int J Robot Autom (IJRA) 2(2):73–82

Faugeras O, Lustman F (1988) Motion and structure from motion in a piecewise planar environment. Int J Pattern Recognit Artif Intell 485–508

Jang JR (1993) ANFIS: adaptive-network-based fuzzy inference system. IEEE Trans Syst Man Cybern 665–685

Kurnaz S, Cetin O, Kaynak O (2010) Adaptive neuro-fuzzy inference system based autonomous control of unmanned air vehicle. Expert Syst Appl 37:1229–1234

Loop C, Zhang Z (1999) Computing rectifying homographies for stereo vision. IEEE Comput Soc Conf Comput Vision Pattern Recognit 1:125–131

Mitra S, Hayashi S (2000) Neuro-fuzzy rule generation: survey in soft computing framework. IEEE Trans Neural Netw 748–768

Nedeljkovic I (2004) Image classification based on fuzzy logic. Int Arch Photogrammetry, Remote Sens Spat Inf Sci 34

Patil MJ, Hodges DH (2006) Flight dynamics of highly flexible flying wings. J Aircr 43(6): 1790–1798

Pitcher NA, Black JT, Reeder MF, Maple RC (2009) Videogrammetry dynamics measurements of a lightweight flexible wing in a wind tunnel. In: 50th AIAA/ASME/ASCE/AHS/ASC structures, structural dynamics, and materials conference, 17th 4–7 May 2009

Rahimi MR, Ghasemi R, Sanaei D (2013) Designing discrete time optimal controller for double inverted pendulum system. J Numer Anal Eng 1(1)

Tamaskani R, Bazzazi A, Mohammadi A, Ajoudani M (2015) Investigating the performance of fuzzy, PID and LQR controllers for control of airplane pitch angle. Int J Nat Eng Sci 9(1): 13–17

Weisshaar TA, Ashley H (1973) Static aeroelasticity and the flying wing. J Aircr 10(10):586–594

Xuebo Z (2008) A fast homography decomposition technique for visual servo of mobile robots. In: Proceedings of the 27th Chinese control conference, Kunming, Yunnan, China, July 2008, pp 16–18

The Dynamics Aspects of Modeling and Control of the Flying Robot with Attached Two Degree of Freedom Manipulator

Grzegorz Chmaj, Karol Seweryn, Tomasz Rybus, Tomasz Buratowski, M. Musioł and Marek Banaszkiewicz

1 Introduction

The UAV is an acronym for Unmanned Aerial Vehicle. UAVs flying autonomously based on the pre-programmed path, or are controlled remotely (by the pilot using a ground control station). Semi-autonomous mode and collision avoidance aided control system are also applied. Initially the main use of UAVs were military missions. Autonomous aircrafts performed intelligence, surveillance, and reconnaissance (ISR) in military tasks. Nowadays their roles have expanded to other areas.

UAV's are equipped with many sensors that gather data from the environment for further use by the control system. The objectives of these control systems are different, starting from locating another aircraft in the space and ending with avoiding collisions. Different solutions with different abilities are given depending on the aircraft mission (Bielecki et al. 2012).

UAVs can be classified by the size from a Micro Air Vehicle weighing one pound to large drones weighing over forty thousand pounds. Examples of different types of UAVs are presented below.

The first example is the advanced solution of UAV, the HALEs (High altitude, long endurance). This aircraft has been designed for obtaining advanced intelligence and reconnaissance work. It has a 46 m wingspan and 200 kg load-carrying capacity. The main characteristic of this UAV is the nearly 20,000 m of service ceiling and ability of flight up to 4 days. Second example is the Nano hummingbird. This semi autonomous UAV was created for surveillance and reconnaissance.

G. Chmaj (✉) · T. Buratowski · M. Musioł
AGH University of Science and Technology, Cracow, Poland
e-mail: grzesiek.chmaj@gmail.com

K. Seweryn · T. Rybus · M. Banaszkiewicz
Space Research Centre of the Polish Academy of Sciences, Warsaw, Poland

J. Sasiadek, *Aerospace Robotics III*, GeoPlanet: Earth and Planetary Sciences,
https://doi.org/10.1007/978-3-319-94517-0_8

The difference with the HALE example is size. This UAV has the form of a real hummingbird and can move at a maximum speed of 11 mph, being able to go forwards, backwards, sideways and rotate on its axis. Opportunities and challenges with autonomous micro aerial vehicles are described in Kumar and Michael (2012). Another interesting examples of UAVs are windows cleaning flying robots (Albers et al. 2010).

As we have seen, there are really different specifications for UAVs depending on the mission and technology. There are many problems to solve e.g. complexity of the cooperation: depending on how strong is the cooperation, faster or slower responses are needed. There are also some barriers that prevent a higher investment, overall in civil applications. The main one is that some kind of UAVs do not have an airspace authorization. But there are also technological problems like the lack of methodology to test the capacity of avoiding collisions (Shima and Rasmussen 2009).

Flying robots with manipulators have been proposed since the 1980s (initially for space applications). Types of robots used in space, their main design features and possible applications, review of space manipulators and space robots is presented in Sasiadek (2013). The dynamics and kinematics of space manipulators systems were described in Vafa and Dubowsky (1987, 1990). A review of dynamic control and modeling for flying robots in space can be found in Moosavian and Papadopoulos (2007). Diverse control system implementation techniques for autonomous robots are described in Zielinski (2000). Nowadays unmanned aerial systems more often are connecting with grasping capabilities. A RC helicopter with attached gripper is described in Pounds and Dollar (2010). The stability of the UAV in contact with the object is described in Pounds et al. (2011). Cooperative grasping capability of quadrotors is presented in Mellinger et al. (2010, 2011). Orsag et al. (2012) describing controlling a mobile manipulating using a commercially available UAV aircraft with three 2 DoFs arms. Kondak (2012) presents a controller for the particular case of a small scaled autonomous helicopter equipped with a robotic arm for aerial manipulation.

In realized project unmanned helicopter was used as a test platform for examination of control algorithms, which could be used on the satellite-manipulator system. Helicopter operating in Earth gravitational conditions cannot be treated as an equivalent of the satellite, although both of these systems have some similarities: the body has 6 Degree of Freedom and the force torque generation creates the disturbances. The influence of the attached manipulator on helicopter may be investigated by observing the state of the body or control signals. However, the system has several differences like additional degrees of freedom in the helicopter case for main and tail rotors (Seweryn et al. 2012).

Introduction of the numerical model of the helicopter and manipulator arm is described in Sect. 2. Section 3 describes in details UAV helicopter which was used in the testing campaign. In Sect. 4 the details of the manipulator specially developed in SRC PAS for testing purposes is presented. The Sect. 5 summarizes the results of 4 campaigns done during last two years. The paper is conclude in Sect. 6 together with remarks about planned future work.

2 Introduction

Research on the UAV with attached manipulator started with the formulation of a numerical model describing the dynamics of a flying robot. At the beginning numerical model of helicopter described in Padfield (2007) was implemented, subsequently taking into account the impact of the installed manipulator. That section is followed by the description of the LQR control system used in simulations and PID control system used on real UAV helicopter.

2.1 Numerical Model of a Helicopter

In our study simple non-linear model presented by Hald et al. (2005) is used to describe the dynamic behavior of a helicopter. In this approach helicopter is considered as a rigid body, free to move and rotate, thus having 6 DoF. Helicopter motion is determined by forces and torques exerted on the helicopter CG (Center of Gravity) by main and tail rotor. The following equations are used for calculation of these forces $\boldsymbol{F}$ and torques $\boldsymbol{T}$:

$$\boldsymbol{F} = \begin{bmatrix} f_x \\ f_y \\ f_z \end{bmatrix} = \begin{bmatrix} -T_{MR}\sin(\beta_{1c}) - \sin(\theta)mg \\ T_{MR}\sin(\beta_{1s}) + T_{TR} + \sin(\phi)\cos(\theta)mg \\ -T_{MR}\cos(\beta_{1s})\cos(\beta_{1c}) + \cos(\phi)\cos(\theta)mg \end{bmatrix} \tag{1}$$

$$\boldsymbol{T} = \begin{bmatrix} L \\ M \\ N \end{bmatrix} = \begin{bmatrix} f_{y,MR}h_m - f_z y_m + f_{y,TR}h_t + Q_{MR}\sin(\beta_{1c}) \\ -f_x h_m - f_z l_m - Q_{MR}\sin(\beta_{1s}) \\ f_x y_m + f_{y,MR}l_m - f_{y,TR}l_t + Q_{MR}\cos(\beta_{1c})\cos(\beta_{1s}) \end{bmatrix} \tag{2}$$

$$\boldsymbol{F} = \begin{bmatrix} f_x \\ f_y \\ f_z \end{bmatrix} = \begin{bmatrix} -T_{MR}\sin(\beta_{1c}) - \sin(\theta)mg \\ T_{MR}\sin(\beta_{1s}) + T_{TR} + \sin(\phi)\cos(\theta)mg \\ -T_{MR}\cos(\beta_{1s})\cos(\beta_{1c}) + \cos(\phi)\cos(\theta)mg \end{bmatrix}$$

where T_{MR} is the main-rotor thrust, T_{TR} is the tail-rotor thrust, β_{1s} is the lateral flapping angle, β_{1c} is the longitudinal flapping angle, θ is the Euler angle of helicopter pitch, φ is the Euler angle of helicopter roll, m is the helicopter mass, g is the gravitational acceleration, l_m, y_m and h_m are the distances between the rotor hub and helicopter CG along the x, y and z axes respectively, l_t is the distance from the center of the tail rotor to CG along z axis, Q_{MR} is the main rotor drag, $f_{y,MR}$ is the force along y caused by the main rotor and $f_{y,TR}$ is the force along caused by the tail rotor. The following equation was used to estimate the main rotor thrust (Heffley and Mnich 1988):

$$T_{MR} = (\omega_b - v_i)\frac{\rho\,\Omega R^2 aBc}{4} \tag{3}$$

where ω_b is the wind velocity relative to the blade, v_i is the induced velocity, ρ is the density of the air, Ω is the rotational velocity of the main-rotor, R is the radius of the blade, a is the two-dimensional constant lift curve slope, B is the number of blades and c is the mean blade cord length The tail-rotor thrust is estimated as follows:

$$T_{TR} = \frac{f_{y,MR} l_m + f_x y_m - Q_{MR}\cos(\beta_{1c})\cos(\beta_{1s})}{l_t} + u_{ped} \tag{4}$$

where u_{ped} is the rudder control input. The main-rotor drag is described by:

$$Q_{MR} = -\left(A_{Q,MR} T_{MR}^{1.5}\right) + B_{Q,MR} \tag{5}$$

where $A_{Q,MR}$ is the relationship between the main-rotor thrust and the drag, while $B_{Q,MR}$ is the initial drag of the main rotor when the blade-pitch angle is zero. The model for the main-rotor flapping used for computation of β_{1c} and β_{1s} is based on the model presented by Mustafic et al. (2005), with simplifying assumption that there is no cross coupling in the main-rotor flapping.

Helicopter is controlled by four control signals. Apart from u_{ped} that was introduced in the Eq. (4) we have: the lateral control input u_{lat}, the longitudinal control input u_{long} and the collective control input u_{col}. Control vector for the helicopter can thus be formulates as follows:

$$\boldsymbol{u_B}(t) = \begin{bmatrix} u_{lat} & u_{long} & u_{col} & u_{ped} \end{bmatrix}^T \tag{6}$$

Components u_{lat}, u_{long} and u_{col} do not appear explicitly in the presented equations. These components are used for controlling the main rotor and are present in equations describing the main rotor flapping (angles β_{1c} and β_{1s}). Detailed description of the approach used to compute all quantities appearing in Eqs. (1)–(5) can be found in (Hald et al. 2005).

For the purpose of numerical simulations the state vector defined as follows was used:

$$\boldsymbol{x} = \begin{bmatrix} \boldsymbol{P} & \boldsymbol{V} & \boldsymbol{\Theta} & \boldsymbol{\omega} \end{bmatrix}^T \tag{7}$$

where $\boldsymbol{P}$ is the position of the helicopter CG, $\boldsymbol{V}$ is its velocity, $\boldsymbol{\Theta}$ denotes helicopter attitude (Euler angles) in respect to the inertial reference frame and $\boldsymbol{\omega}$ is the helicopter angular velocity projected onto the body frame. It should be noted that the flapping angles β_{1s} and β_{1c} are not included in the state vector $\boldsymbol{x}$ and must be threaded separately in the modeling (Hald et al. 2005). However, state vector (7) can be expended to include additional state variables, e.g., state of manipulator joints in case of the helicopter equipped with a manipulator (Sect. 2.2). Motion of the helicopter is described by the rigid body equations that can be expressed in the matrix form (8):

$$\dot{x} = \begin{bmatrix} \dot{\boldsymbol{P}} \\ \dot{\boldsymbol{V}} \\ \dot{\boldsymbol{\Theta}} \\ \dot{\boldsymbol{\omega}} \end{bmatrix} = \begin{bmatrix} \boldsymbol{V} \\ \frac{1}{m}\boldsymbol{F} - \boldsymbol{\omega} \times \boldsymbol{V} \\ P_{sb}(\boldsymbol{\Theta}) \cdot \boldsymbol{\omega} \\ \boldsymbol{I}^{-1}(\boldsymbol{T} - \boldsymbol{\omega} \times (\boldsymbol{I} \cdot \boldsymbol{\omega})) \end{bmatrix} \tag{8}$$

$$\dot{\boldsymbol{\Theta}} = P_{sb}(\boldsymbol{\Theta}) \cdot \boldsymbol{\omega} \tag{9}$$

where $\boldsymbol{I}$ is the inertia tensor of the helicopter and $P_{sb}(\boldsymbol{\Theta})$ is the transformation matrix that maps angular velocities between the body reference frame and inertial reference frame through the relation (9).

2.2 Manipulator—UAV Coupled Dynamic Model

The manipulator used on the UAV system consists of two kinematic pairs: rotational and translational. The kinematic scheme of this manipulator is presented in the (Fig. 1).

Developed numerical model of the helicopter has been expanded with the manipulator dynamics equations. As in many earlier studies, to deal with the problem, the Lagrangian formalism was selected. The generalized coordinates described by Eq. 7 was extended to the following form where q = $[x\ \alpha]$, where x represent the helicopter state (Eq. 7) and additional vector $\alpha = [q\ dq]$ in coupled system represents joint variable of the manipulator arm. The second order Lagrangian equations (version for quasi coordinates) was used to derive the generalized equations of motion:

$$\begin{bmatrix} M_B & M_{BM} \\ M_{BM} & M_M \end{bmatrix} \begin{bmatrix} \ddot{x} \\ \ddot{\alpha} \end{bmatrix} + \begin{bmatrix} C_B & C_{BM} \\ C_{BM} & C_M \end{bmatrix} \begin{bmatrix} \dot{x} \\ \dot{\alpha} \end{bmatrix} + G = \begin{bmatrix} F \\ T \\ \tau \end{bmatrix} \tag{10}$$

where B, M, BM subscripts of the mass M and Coriolis C submatrixes indicates the body of helicopter, the body of manipulators and mix terms BM, respectively. The G is the part of equation connected with gravity potential. The right side parts of Eq. 10 represents forces F, torques T acting on helicopter described by Eq. 1

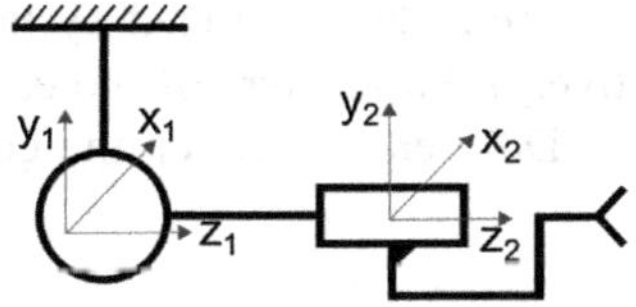

Fig. 1 The kinematic scheme of the manipulator

and 2. These variables cover control system described in Sect. 2.3. The term τ is responsible for manipulator actuation and it is equal to the input $\tau = u_M$.

2.3 Flying Robot Motion Control Systems

Control of a helicopter equipped with a manipulator is a challenging task, as changes in manipulator configuration will results in significant changes of parameters of the controlled system (e.g., position of helicopter-manipulator CG). Moreover, motion of the manipulator can cause disturbances, e.g., due to oscillations of the lightweight manipulator. Thus, control system with high resistance to disturbances and ability to adapt for changing parameters of the object is required.

UAV helicopters used for tests were equipped with PID controller. This controller was manually adjusted before flight to take into account additional mass of the manipulator. However, it was not possible to adjust parameters of the PID control during the flight in response to changes of the manipulator configuration. As structure of the PID controller used in the UAV helicopter was not known, the model of the controller could not be implemented in the numerical simulations. It was decided to perform numerical simulations with LQR controller instead of simple PID. LQR controller works better in forcing simple numerical model of the helicopter to follow predefined path. Thus, LQR controller used in simulations allowed to obtain results closer to the reality.

Stationary LQR control method, that is considered in our study, is one of the optimal control methods (McLean 1991). These methods are based on the synthesis of control laws in such a way that the selected cost functional of control quality has a minimum value (optimal control theory is an extension of the calculus of variations). Such optimization methods allow synthesis of control, which is less sensitive to fluctuations of parameters of object and regulator, than the traditional methods used in the design of automation systems.

In the case of LQR control method, non-linear model of the object (helicopter) must be linearized around the point of equilibrium. Such description will be valid as long as the object state is close enough to the state of equilibrium for which linearization was made. Linearized model has the form:

$$\dot{\boldsymbol{x}}(t) = \dot{\boldsymbol{x}}_0 + \boldsymbol{A}(\boldsymbol{x}(t) - \boldsymbol{x}_0) + \boldsymbol{B}(\boldsymbol{u}_B(t) - \boldsymbol{u}_{B0}) \tag{11}$$

where $\boldsymbol{x}(t)$ is the state vector that contains 12 variables describing state of the helicopter (defined in the same way as in the Sect. 2.1), $\boldsymbol{u}_B(t)$ is the helicopter control vector (defined in Eq. (6)), $\boldsymbol{A}$ and $\boldsymbol{B}$ are time-independent matrices that describe the linearized system (having dimensions 12×12 and 12×4 respectively). Parameters of the equilibrium state are denoted by the subscript 0.

Difference between the current state and reference state is expressed as:

$$\boldsymbol{x}_s(t) = \boldsymbol{x}(t) - \boldsymbol{x}_{ref}(t) \tag{12}$$

and task of the control system is to bring $\boldsymbol{x}_s$ to zero. Cost function is assumed in the form of a quadratic functional of control quality:

$$J = \frac{1}{2}\int_0^{\infty} \left[\boldsymbol{x}_s^T(t)\boldsymbol{Q}\boldsymbol{x}_s(t) + \Delta\boldsymbol{u}_B^T(t)\boldsymbol{R}\Delta\boldsymbol{u}_B(t)\right]dt \tag{13}$$

where $\boldsymbol{Q}$ is 12 × 12 time-independent non-negatively defined weighting matrix for the output vector, while $\boldsymbol{R}$ is 4 × 4 time-independent positively defined weighting matrix for the control vector and:

$$\Delta\boldsymbol{u}_B(t) = \boldsymbol{u}_B(t) - \boldsymbol{u}_{B0} \tag{14}$$

Matrices $\boldsymbol{Q}$ and $\boldsymbol{R}$ are selected basing on the results of numerical simulations, knowledge of the object and its expected behavior in response to the differences between the current and reference state vector. In considered case control system is required to move helicopter along the reference trajectory defined as the time function of position (x and y component), height and yaw angle. Matrix $\boldsymbol{Q}$ can be selected in such a way, that only differences between these four components of current and reference state vector are taken into account in computation of the control vector $\boldsymbol{u}(t)$.

For computation of the control vector $\boldsymbol{u}_B(t)$ the following control law is used:

$$\boldsymbol{u}_B(t) = \boldsymbol{K}_B\boldsymbol{x}_s(t) \tag{15}$$

Time-independent gain matrix $\boldsymbol{K}_B$ is computed from the equation:

$$\boldsymbol{K}_B = \boldsymbol{R}^{-1}\boldsymbol{B}^T\boldsymbol{P} \tag{16}$$

where matrix $\boldsymbol{P}$ is the solution of the algebraic Riccati equation:

$$\boldsymbol{PA} + \boldsymbol{A}^T\boldsymbol{P} - \boldsymbol{PBR}^{-1}\boldsymbol{B}^T\boldsymbol{P} + \boldsymbol{Q} = 0 \tag{17}$$

In the presented approach weighting matrices $\boldsymbol{Q}$ and $\boldsymbol{R}$ are constant and information on the state of the manipulator is not used by the control system. Nevertheless, LQR control method should not be very sensitive to slow changes of parameters of controlled object. Therefore, assuming that the motion of the manipulator is relatively slow, control law (15) with constant gain matrix $\boldsymbol{K}_B$ should be sufficient for controlling the helicopter.

Second control system described in this paper is based on the Proportional Integral-Differentiational (PID) controllers that are part of the original autopilot of the Aquila helicopter used for test flights. Helicopter operator has possibility to make changes in the settings of the PID controller, depending on the conditions and the parameters of the helicopter (e.g., additional weight carried by the helicopter). Adjustment of PID settings could also be performed automatically (basing on the predefined tables of coefficients) in response to changes of the manipulator configuration. Thus, data from manipulator encoders could be used by the control system to enhance its performance.

Separate issue is the control of the manipulator. As 2 DoF manipulator was used, control vector for the manipulator has two components (18).

$$\boldsymbol{u}_M(t) = \begin{bmatrix} u_{J1} & u_{J2} \end{bmatrix}^T \tag{18}$$

$$\boldsymbol{u}_M(t) = \boldsymbol{K}_M \boldsymbol{\alpha}_{Ms}(t) \tag{19}$$

In numerical simulation LQR controller was also used for control of the manipulator. Approach for computation of the gain matrix K for the manipulator is the same as in the case of the LQR controller for the helicopter. The only difference is in the dimensions of matrices appearing in equations presented in this section. In this case Eq. (15) can be replaced with Eq. (19) where α_{Ms} is the difference between the current state and reference state of the manipulator.

During the test-flights, PID controllers were used to control the manipulator joints and signal from encoders were used to compute difference between current and planned position of joints.

2.4 Simulations Tools

The modeling and simulation of a helicopter with onboard manipulator is a challenging task since significant interactions between separate components of such system occur. It is necessary to take into account dynamic interactions (e.g. motions of the manipulator induce reaction torques and reaction forces—on the object to which the manipulator is attached) and interactions between control systems (e.g. manipulator's joint controller acts independently of the controller responsible for keeping desired attitude of the object). In case of the helicopter, modeling interactions caused by the aerodynamic forces acting on the rotor must also be considered (Padfield 2007). Object oriented modeling is especially suitable for simulations in aerospace robotics because it allows straight forward combination of various subsystems and environments (Ferretti et al. 2004). Simulations of helicopter equipped with a manipulator described in this paper were performed with specific simulation tools developed in the Space Research Centre PAS. This tools are based on the models built in SimMechanics (software based on the Simscape, the platform product for the MATLAB Simulink). Simscape solver works autonomously and creates the physical network from the block model of the system and then the system of equations for the model is constructed and solved (Haug 1989) and (Wood 2003). The use of this platform allowed incorporation into one environment all the elements essential in numerical simulations of complex

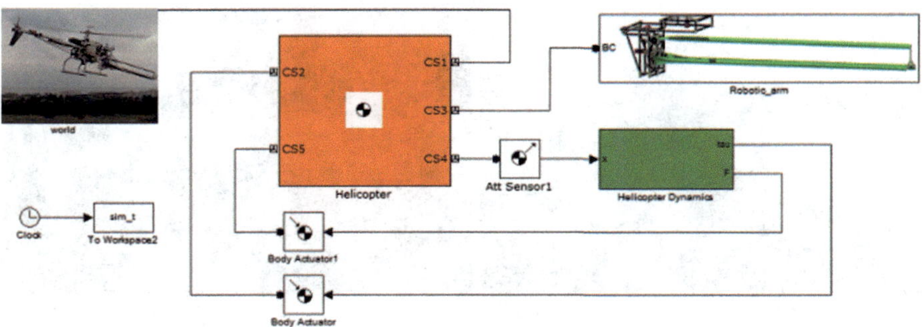

Fig. 2 UAV helicopter dynamics model

satellite-manipulator and helicopter-manipulator systems. The helicopter is modeled with the equations of helicopter dynamics described in Sect. 2.1. The implementation of these equations is based on the description provided by Hald (2005). Equations of the helicopter dynamics are incorporated into Simulink model as Embedded Matlab Functions. Forces and torques computed with the Eqs. (1) and (2) are then applied to rigid body modeled with SimMechanics. The manipulator modeled in SimMechanics consists of two kinematic pairs: one rotational and one translational.

LQR controller is responsible for generating control signals (Eqs. 6 and 15) for helicopter's trajectory following predefined path. This controller will also counteract disturbances generated by the motion of the manipulator. The manipulator arm motion is controlled also by LQR (Eq. 19) which is only responsible for assuring realization of selected manipulator's trajectory in manipulator frame.

The simulation tool presented here is based on space robot simulator developed in SRC PAS (Seweryn et al. 2011), updated to be able to simulate UAV helicopter with manipulator. The tool itself described in this section was previously published in our previous paper (Seweryn et al. 2012) (Fig. 2).

3 Description and Parameters of Analyzed Systems

Two different unmanned helicopters were used for test-flights. Preliminary tests and first verification of the numerical model of the helicopter dynamics were performed on the IRL (Chmaj 2010). Before the final design and assembly of the tubular-beam manipulator this helicopter was also used for test-flights with mockup of the manipulator.

Such tests were conducted in order to determine the levels of vibrations on the tubular-beams mounted on a helicopter. Tests with the final version of the manipulator were performed on the Aquila helicopter, which is bigger than IRL

Fig. 3 Unmanned helicopters used for tests: IRL (left) and Aquila (right)

Table 1 Parameters of helicopters used in tests

Parameter	IRL	Aquila
Number of blades	2	2
Main rotor angular velocity (rad/s)	140	150
Main rotor diameter (m)	1.6	2.134
Mean blade cord length (m)	0.06	0.07
Instrumented vehicle weight + Fuel (kg)	9	24
Inertia IXX (kg·m^2)	~0.18	~0.89
Inertia IYY(kg·m^2)	~0.76	~3.66
Inertia IZZ (kg·m^2)	~0.72	~3.37

(24 kg of the instrumented vehicle weight in comparison to 9 kg for IRL). Pictures of both helicopters are presented in the Fig. 3 and basic parameters of helicopters can be found in the Table 1.

3.1 Onboard Flight Control System

Helicopter flight control system is a combination of sensors, GNSS (Global Navigation Satellite System), software and control system which regulate aircraft actuators. Regardless of the system purpose (F-16 fighter, or a small unmanned aircraft UAV), the primary system task is estimate of: position (latitude, longitude, altitude), velocity (speed in the direction of north, east, vertical) orientation (direction of flight, pitch and roll of the aircraft). To make estimation of these parameters possible, integration of several smaller systems providing information about mentioned parameters should be done. These are:

- Inertial Measurement Unit system

IMU is a combination of gyro sensors and accelerometers to measure angular velocity and acceleration around the axes X, Y, Z, and along the X, Y, Z,

- System AHRS (Attitude and Heading called Reference System). AHRS is a combination of IMU system with the Earth's magnetic field sensor that provides information about the direction of flight,

- GNSS system provides information about the location, speed and flight altitude.

The use of GNSS in the INS system can be implemented in two configurations: a loosely coupled and tightly coupled. For loosely coupled configuration, when the vehicle is in contact with less than four satellites, the most common Kalman filter (KF) is not able to estimate a fairly good estimate of the position and speed of the vehicle. For tightly coupled configuration, the Kalman filter takes into account data even from a single satellite.

The flight control systems can be categorized in two groups: the flight control systems operating in open-loop and flight control systems operating in closed loop. The first group includes civilian use, where the final decision is made by the pilot. In the second group are military applications and UAVs, for which the final decision is always made by the flight control system.

IRL used for research, has been equipped with INS system built from commercially available components. Its main features are: the possibility of vertical take-off maneuver, the ability to program its own control algorithms, communication with terrestrial base station, full configurability of all system parameters.

The main component of the system is the MP2128Heli system that has been equipped with a number of components such as the Earth's magnetic field sensor, ultrasonic sensor used during VTOL maneuver, eight-channel analog digital converter (ADC) allows connection of additional external sensors (e.g. to measure the fuel level in the tank, engine temperature and engine speed) and radio modem. The heart of the MP2128Heli system is a microprocessor with efficiency of 150 million instructions per second (150 MIPS), made by RISC company. MP2128Heli has a built-in IMU 4P TIM GPS receiver, working at a frequency of 4 Hz. Information about the orientation of the helicopter is provided to MP2128Heli on the basis of IMU system. The primary source of information about robot's location is a GPS receiver. Algorithms running on the chip contain two 6-state Kalman filters. These filters operate in a real time with signal sampling rate of 200 Hz. One of them is integrated with the GPS receiver in the loosely coupled configuration. System weighs only 28 g and its dimensions are 100 × 40 × 20 mm. In INS/GPS systems. Kalman filter is used to estimate errors connected with estimates of position, velocity and orientation. Estimation errors are results of the sensor's work nature, sensors quality or the system's nature. In commercial systems INS/GPS, even 50 state vector variables are estimated. Note, however, that the computational power needed to use the Kalman filter (KF) is the third power of the number of state variables. The values of the estimated states are as accurate, as the mathematical model of the system's dynamic, that is being run by the KF. Therefore, when computational power is limited, the number of the state variables is in the range of 12–17. It should be noted, that for 17 state vector variables about 15 variables are errors of the IMU, and only two or three state variables are errors connected with the GPS system. The reason is the multi-sensor construction of IMU. Typical errors of the IMU and GPS are shown in Table 2 (Fig. 4).

Table 2 The of the IMU and GPS main error sources

Gyro errors	Accelerometer errors	GPS errors
Nonlinearity	Squaring	Ionospheric delay
–	Weight imbalances	
Bias caused by temperature		Tropospheric delay
Scale factor error		Ephemeris error
Asymmetry of scale factor error		Satellite clock error
Nonorthogonal axis sensor errors		Multipath effect
White noise		Receiver errors
Correlated noise		
Correlation time		
Bias		

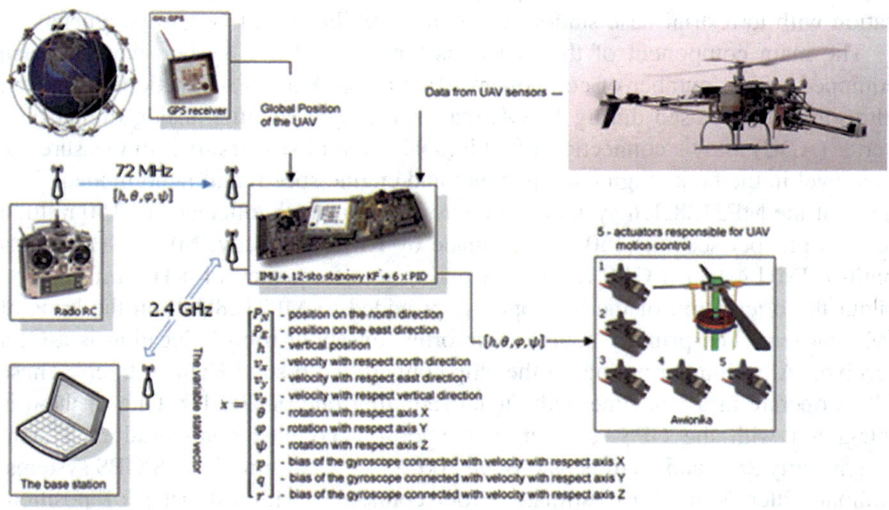

Fig. 4 Infrastructure-board flight control system and its components

MP2128Heli system use two Kalman filters responsible for estimating a states related to the orientation and position of the robot. Vectors state of these filters are shown below. State vector that store information about the orientation of the robot:

$$[\theta, \varphi, \psi, \theta_{bias}, \varphi_{bias}, \psi_{bias}] \tag{20}$$

where:

θ rotation around the X axis
φ rotation around the Y axis
ψ rotation around the Z axis

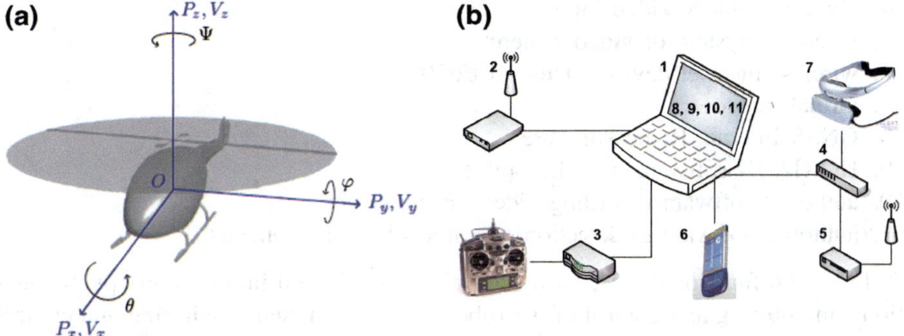

Fig. 5 **a** Coordinate system associated with the robot. **b** Elements of the flight control station

θ_{bias} systematic error of gyro measuring the angular velocity around the axis X
φ_{bias} systematic error of gyro measuring the angular velocity around the axis Y
ψ_{bias} systematic error of gyro measuring the angular velocity around the axis Z.

State vector that contains information about the position of the robot:

$$\left[P_x, P_y, P_z, V_x, V_y, V_z\right] \tag{21}$$

where:

P_x position in a northern direction (X direction)
P_y position in an eastern direction (Y direction)
P_z position in the vertical direction (Z direction)
V_x velocity in a northern direction (X direction)
V_y velocity in a eastern direction (X direction)
V_z velocity in the vertical direction (Z direction).

The X, Y, Z axes in descriptions of Eqs. (20) and (21) form the axes of Cartesian coordinate system associated with the robot shown in Fig. 5a.

3.2 Terrestrial Flight Control Station

Ground flight control station is a combination of computer integrated with different components (supporting the mission in which the robot is involved) and software (Fig. 5b). The station consists of:

1. computer
2. MHX-2400's radio modem (by Microhard Systems)
3. high-frequency module control system for Optic 6 radio (by Hitec)

4. TV card with S-Video input
5. receiving system of video system
6. wireless internet device—Huawei E620
7. virtual glasses
8. GNSS Internet Radio software
9. HORIZON software (by Micropilot)
10. authorial software recording video on the server
11. authorial software to detection operator's head movements.

From the functional side, base station can be divided into two groups of functions: monitoring and control of the robot's flight data, and monitoring and control of the robot's additional equipment needed to accomplish the mission. The first functional group, almost entirely, is supported by HORIZON software. Horizon allows for example loading maps of the area where the robot mission will be carried out. As a result, the current flight path can be seen on the map. Another task of the software is to monitor the current state of flight parameters from telemetry data.

4 Manipulator

Analysis of possible design solutions of manipulator was based on the technical requirements for this system. It was assumed that the mass of the manipulator must be less than 4 kg (this is due to the maximum capacity of helicopter), manipulator should have no resonance frequencies around the main resonant frequency of the helicopter (180 and 25 Hz) and the maximum length of the manipulator should be not less than 1.4 m.

Two concepts of manipulator were considered: the first one based on truss system and the second one based on tubular-beam solution. For the project needs a tubular-beam manipulator was chosen (Kuciński et al. 2014).

The concept of tubular-beam manipulator is based on a system with two degrees of freedom. The first degree allows rotation of the manipulator in the range of 0°–270° while the second allows translational (increasing working space of manipulator outside the line of the rotor blades—0.9 m). Helicopter equipped with a manipulator is shown in Fig. 6.

From the structural side, manipulator rotational DoF can be described as rotation around the vertical axis (rotational kinematic pair). The translation DoF was realized by three steel tapes. The DC motors were used to assert the movement.

The construction is characterized by three factors: very low weight (about 3.7 kg), long-range work (about 2 m), a small volume of the manipulator (dimensions 500 × 300 × 300 mm). Flexible tape technology was tested in several space experiments. Table 3 contains specifics of manipulator. Prepared concept of manipulator was analyzed in FEM software. It determined the optimal arrangement of flexible tapes for which resonant frequencies were appointed. In this case solutions widely used during the construction of manipulators for space

Fig. 6 A helicopter equipped with a tubular-beam manipulator

Table 3 Specification of used manipulator

	Tubular-beam system
Mass	3.7 kg
The speed of the individual members of manipulator	1 m/min
Power consumption	40 W
Dimensions (length x height x width) in mm (composite manipulator)	520 × 320 × 288
Dimensions (length x height x width) in mm (distributed manipulator)	1920 × 320 × 288
Working member permissible load (which can carry the weight of the manipulator)	300 g
Resonant frequency (first 3)	7.76, 12.3, 20.7 Hz
Maintain a constant center of gravity/inertia	Yes/no
Number of degrees of freedom	2
Workspace	Limited plane

applications have been used. Materials such as titanium IMI 318, carbon fiber tubes, aluminum PA9 or plastic Delrin, PEEK have been applied.

5 Prototype Testing

The following chapter summarizes the results of the test flights and the results of numerical simulations.

5.1 Verification of the Helicopter Numerical Model

Hald et al. (2005) performed qualitative comparison of a real helicopter movement and the behavior of the non-linear model he presented. It was shown that the behavior of the non-linear model initialized in hover approximates that of a real helicopter. Seweryn et al. (2012) presented comparison between the measurements obtained during test flights of IRL helicopter and the results of simulations based on the numerical model of a helicopter (described in the Sect. 2.1) that was controlled by the LQR controller (described in the Sect. 2.3). High correlation between helicopter trajectory obtained from measurements and simulations suggests that the numerical model is correct. Relatively good quantitative correlation between the simulated and measured components of the control signal was also obtained. However, it must be noted that in the presented case comparison of the dynamical behavior of the helicopter is not possible, as helicopter control system (real and simulated) was forcing helicopter to move along the reference trajectory. Selected model of helicopter dynamics is too simple to fully resemble the behavior of the real helicopter. If the control signals obtained from the real autopilot are used in simulations, state of the simulated helicopter differs greatly from the measured state of the real helicopter after only a few seconds. A good fit between simulation results and measurements results from the fact that both autopilots (simulated and real) were working in closed-loops for the following of the reference trajectory. In order to obtain more realistic responses of simulated helicopter, more complex modeling must be applied (Bhandari et al. 2005).

In this section selected results of comparison between the measurements obtained during test flights of IRL helicopter and the results of numerical simulations are presented. All these results were previously shown by Seweryn et al. (2012). Altitude of the helicopter is presented in the Fig. 7a, while helicopter heading (yaw angle) is presented in the Fig. 7b. Figure 8 shows helicopter collective control input.

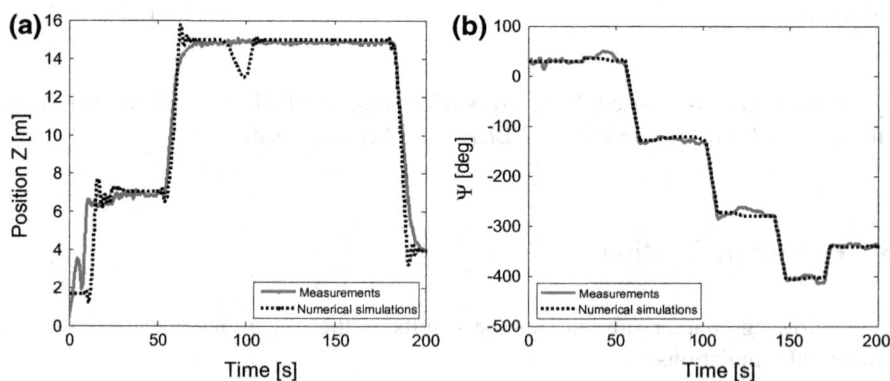

Fig. 7 **a** Helicopter height—comparison between measurements and results of numerical simulations. **b** Helicopter yaw angle—comparison between measurements and results of numerical simulations

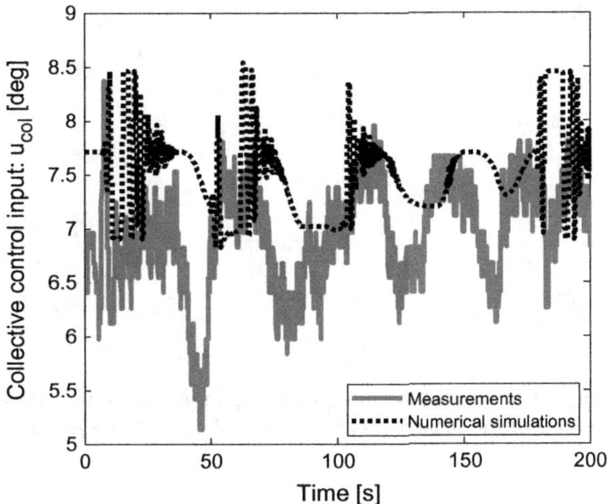

Fig. 8 Helicopter collective control input—comparison between measurements and results of numerical simulations

5.2 *Simulation Verification of LQR Controller*

To test the developed control system for helicopter equipped with a manipulator a simplified simulator was used. The control system had the task to maintain a constant position and orientation of the helicopter when the manipulator trajectory was realized. Figure 9a presents recorded angle of the manipulator, while Fig. 9b presents recorded eject of the manipulator. The test consisted of the following stages: ejecting manipulator at the distance 0.6 m, rotating the manipulator to the left at 75°, rotating the manipulator to the right at 150°, rotating the manipulator to the left at 75° (return to the starting position). Position and orientation of the tip of the manipulator in the inertial coordinate system is shown in Fig. 10a, b, respectively. Figure 11a shows the position of the helicopter in the X-axis, while Fig. 11b

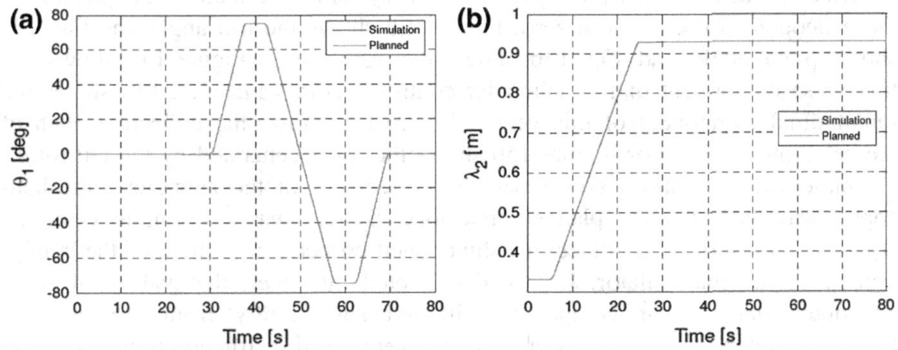

Fig. 9 **a** Angle of rotation around the vertical axis. Simulation results. **b** The ejecting of manipulator. Simulation results

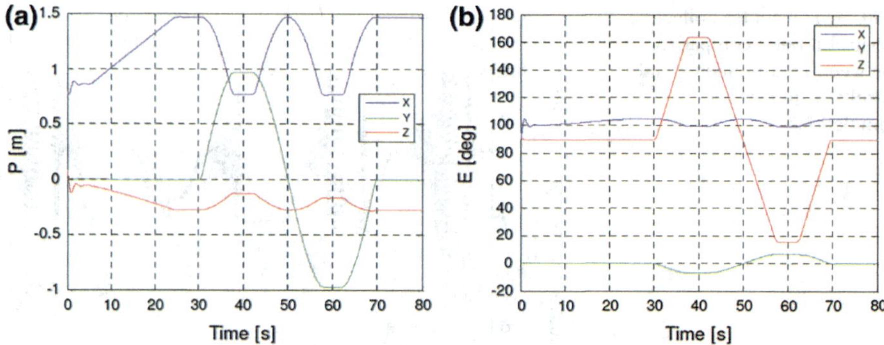

Fig. 10 **a** State of the manipulator tip in inertial coordinate system: position. Simulation results. **b** State of the manipulator tip in inertial coordinate system: orientation. Simulation results

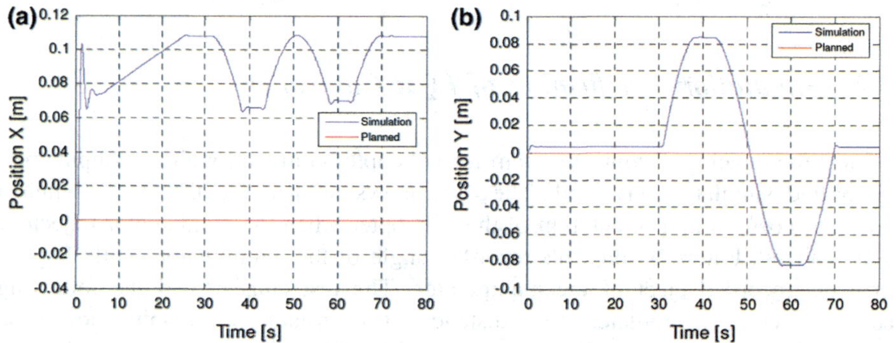

Fig. 11 **a** UAV position in the X axis. Simulation results. **b** UAV position in the Y axis. Simulation results

shows the position of the helicopter in the Y-axis. The position of the helicopter in the Z axis (altitude) is located on Fig. 12a. The system shall maintain the position of the helicopter in each axis at zero. Figure 12b shows the roll angle, the Fig. 13a shows pitch angle and Fig. 13b shows the yaw angle. Figure 14a shows the transverse component of the helicopter control vector, while Fig. 14b shows the longitudinal component of this vector. Figure 15a shows the collective pitch of the helicopter main rotor. Figure 15b shows the force generated by the tail rotor.

Numerical simulations conformed the correctness of the approach—the helicopter was moved along planned trajectory. System was resistant to changing control parameters of the object and functioned properly regardless of the configuration of the manipulator. It should be noted, however, that values of some helicopter state vector components differed significantly from the planned. Especially, pitch and roll angles of the helicopter caused by rotation of manipulator, are twice as high as its suspected values. Pitch angle shown in Fig. 20 is larger than

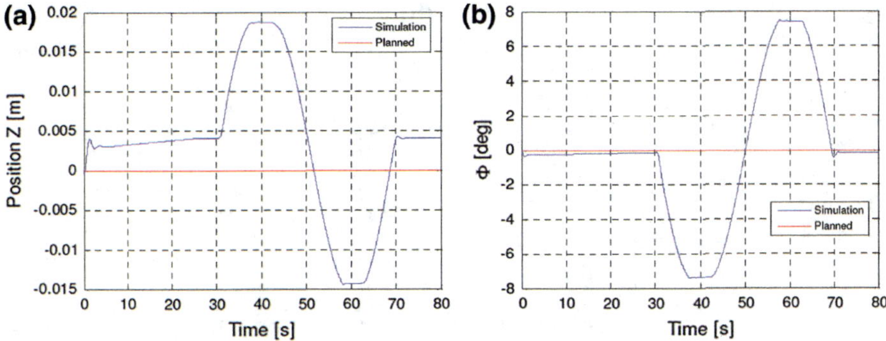

Fig. 12 **a** UAV position in the Z-axis. Simulation results. **b** Roll angle φ. Simulation results

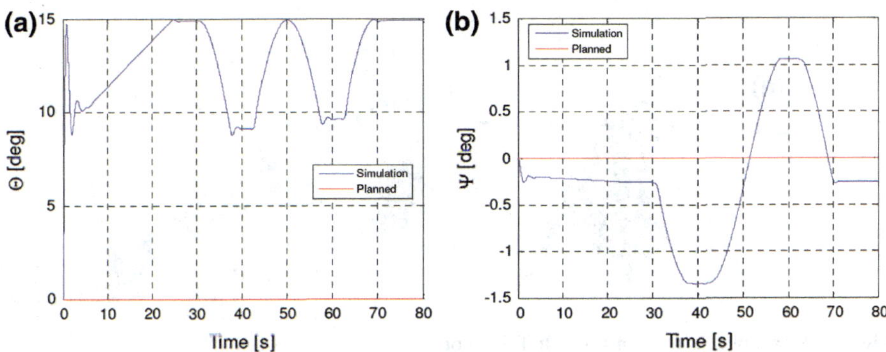

Fig. 13 **a** Pitch angle θ. Simulation results. **b** Yaw angle ψ. Simulation results

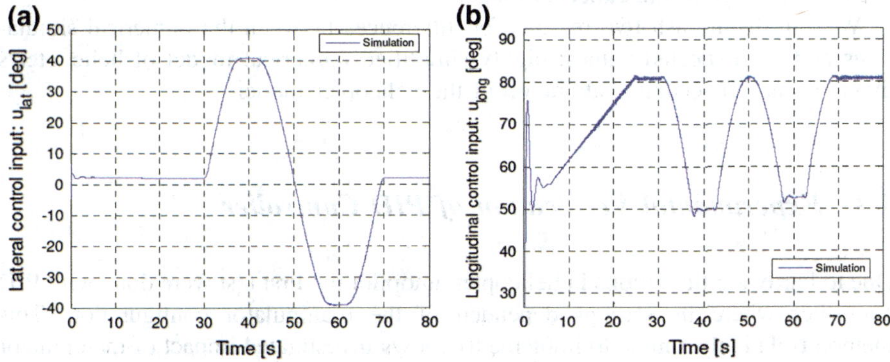

Fig. 14 **a** The transverse component of the helicopter control vector. Simulation results. **b** The longitudinal component of the helicopter control vector. Simulation results

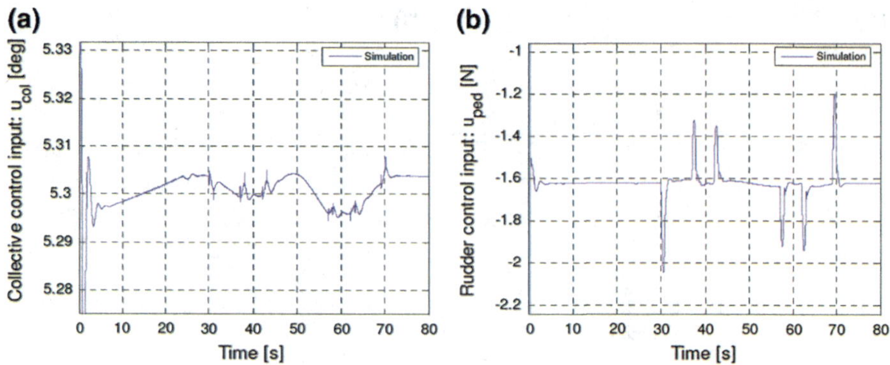

Fig. 15 **a** The collective pitch of the helicopter. Simulation results. **b** The strength of the helicopter tail rotor. Simulation results

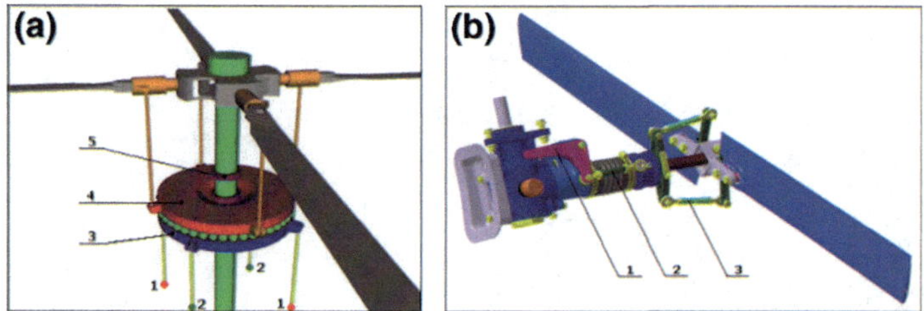

Fig. 16 Swashplates. **a** Main rotor. **b** Tail rotor

planned by 15°. In addition, the values obtained for components of the control vector were beyond the expected range.

We can distinguish two reasons for differences between the numerical simulations and the expected values: highly simplified numerical model of helicopter's dynamics and inaccurate calibration of the helicopter model.

5.3 *Experimental Verification of PID Controller*

Due to hardware limitations in helicopter autopilot the first test were done with PID controller where its settings depended of the manipulator configuration. This method (relatively simple to implement) allows investigated impact of manipulator configuration to the behavior of the helicopter.

In Aquilla helicopter control system with six PID regulators was used. Regulators control the main rotor swashplate actuator (Fig. 16a), the tail rotor swashplate actuator (Fig. 16b) and the carburetor throttle actuator.

Table 4 Used PID regulators

Lp	Regulator name	Regulator description
1	Aileron from roll	Control the aileron deflection. The difference between the specified roll and roll estimate by KF should be minimal
2	Elevator from pitch	Control the elevator pitch. The difference between the specified pitch and pitch estimate by KF should be minimal
3	Hover rudder from heading	Control the general pitch of the tail rotor swashplate. The difference between the specified direction and the direction of flight by KF estimate should be minimal

During the tests, not all regulators are used. Regulators activity depends on the configured type of control. Control loops used during the test are shown in Table 4.

The main objective of the tests was to study the impact of the manipulator movements to the UAV orientation and position. Similarly to real helicopters, hover and flight requires deflection of the throttle and the pitch of the main rotor swashplate. However, with the increase of the main rotor rotations, torque is also increased. Moment must be compensated by the tail rotor in order to maintain a constant direction. Every change of rotor rotations, forces the torque compensation by the tail rotor. This distorts real impact of the manipulator motion on the helicopter orientation. The above-mentioned control is implemented by using so-called lookup tables (also called TLU tables). In order to eliminate discussed disturbance, two tables in the autopilot system were used:

- The first table stores mapping values of throttle deflection on the physical location of the servo controlling the physical deflection of throttle
- Second table stores mapping values of throttle deflection on the physical location of the servo controlling the overall pitch of the main rotor swashplate.

The first table was determined experimentally during helicopter hovering. The slope of first part of the characteristic (Fig. 17a) corresponds to the engine start phase. In this way, going through all the resonances of the robot and manipulator is quick. The value of the flat part of characteristics was chosen to ensure an adequate reserve of drive power required to the lifting and helicopter flight. This part of the characteristics guarantee constant engine speed independently of flight altitude. Finally, the third part of the characteristics is to increasing the engine speed to the maximum value. This part is not use during correct helicopter operations. It provides a power reserve that can be used in emergency situations such as rapid helicopter descent.

The second table has also been set up experimentally. Figure 17b shows a linear characteristic of throttle. This corresponds to a linear change of the angle of attack of the main rotor blades, from negative values to positive values. Configured values of the throttle and collective pitch, allowed to maintain constant speed of helicopter drive during the tests. This allowed to study real impact of the manipulator motion to maintain the orientation of the helicopter.

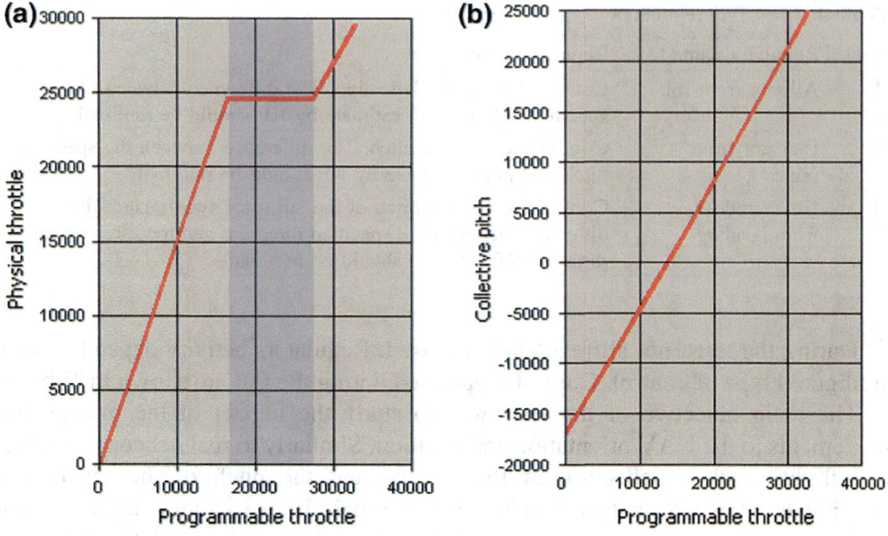

Fig. 17 **a** Graph of deflection actual engine carburetor throttle relative to the program value of throttle. **b** Graph of general pitch of main rotor swashplate relative to the program value of throttle

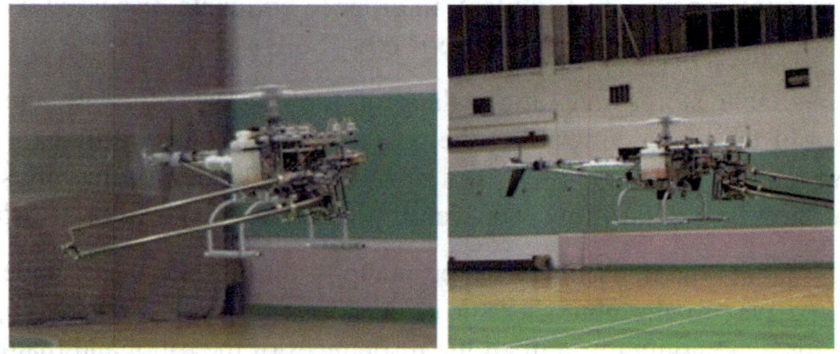

Fig. 18 Manipulator with a UAV helicopter during the functional tests

Figure 18 shows two extreme positions of the manipulator during a helicopter test flights.

The first flights with manipulator mounted on UAV, were aimed to investigate the dynamics of flight during the work of manipulator in its extreme positions. It should be noted that the compensation roll or pitch of the helicopter is done by moving the helicopter's swashplate while the manipulator is using. The swashplate has a some range of motion. Excessive manipulator ejection (and thus change the center of gravity of the object) may causes problems with keeping the robot in the unchanging position. In order to verify the correctness of the adopted manipulator parameters a series of tests were carried out. Figure 19a shows the percentages deflections of individual swashplate servos. The curve in the diagram described as

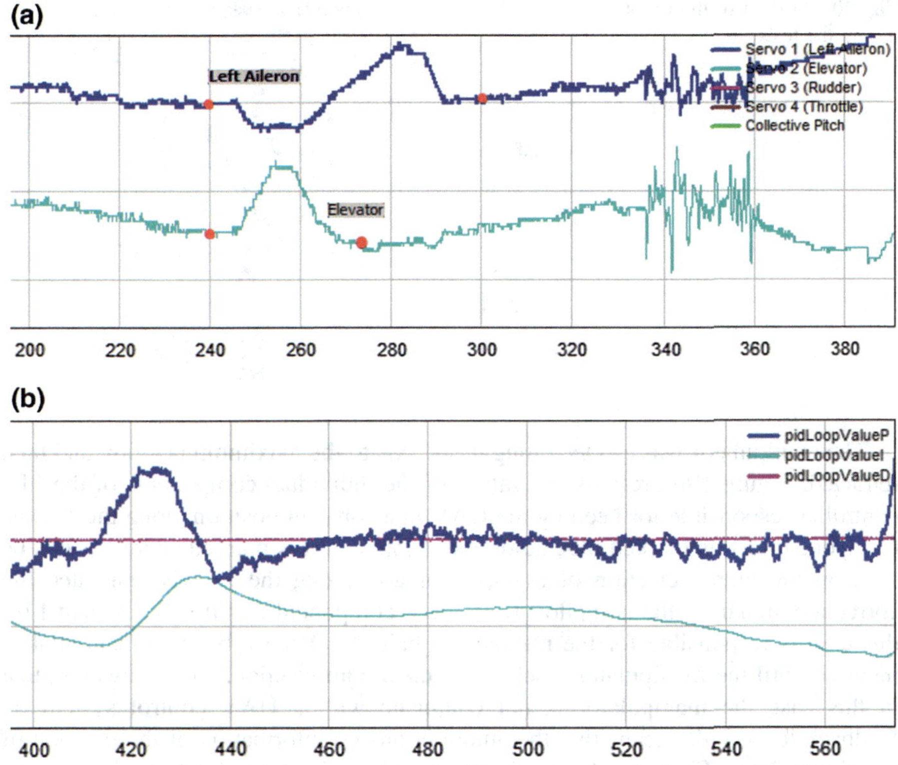

Fig. 19 **a** Percentages deflections of individual swashplate servos. **b** The values of the individual components of the PID controller

"Elevator" corresponds to a situation in which the manipulator is fully ejected and then retracted along the axis X. Servo starting value was 4% and at full ejection of manipulator about 16%. It should be noted that the maximum tilt range is 50% (−50 to 50%—a full range of servo). Curve "Left Aileron" illustrates the changes in value of helicopter roll servo. In this case, the range of motion was between 12 and 2%. The conclusions of the tests were as follows:

- maintain a constant position of the helicopter, even at maximum manipulator arms positions is possible (for the applied weight)
- range of motion necessary to maintain correct servo position (when the manipulator is working) was about 15%. There is a large reserve of compensatory movements allowing attachment of additional weight at the manipulator.

Next objective was to investigate the effect of additional mass attached on manipulator for a helicopter control system. Tests with additional mass was performed in field conditions due to the availability of the GPS signal. UAV maintains its position in an autonomous manner (autopilot), without operator intervention.

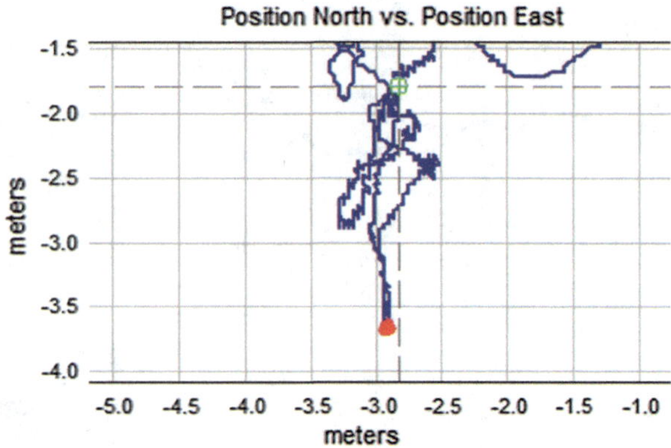

Fig. 20 Position of the robot during the tests

The manipulator was ejected along the X axis to the maximum position and then retracted. Figure 19b presents the values of the individual components of the PID controller responsible for keeping the UAV in a constant position along the X axis. Component of P is marked blue color and component I is marked azure. As can be seen on the graph, ejection of manipulator arm along the X axis generated the correction signal by the autopilot. Both the P component and the component I (in this case is responsible for the mechanical balance of the robot and manipulator) increase until the manipulator reached the maximum ejection. Note, however, that in this case the manipulator is not conjugate with a UAV control system by feedback loop. This means that the autopilot has no information of the position of the manipulator. Ejection of actuator is treated by the autopilot as typical disturbance such as a gust of wind, which changes the position of the robot. In this test the autopilot task was to maintain an unchanging position. Position signal correction along the X-axis has been generated because of the manipulator motion along the x-axis. This situation is depicted in the figure below. Initial robot position (manipulator arm retracted) is marked green, and farthest position of the robot is marked red. For this position correction of autopilot had become so strong that it made it impossible to continue tests. This resulted in a return flight to achieve the desired starting position. As can be seen on the Fig. 20, position error along the X axis was about 3 m. This is the level of accuracy of the GPS receiver.

6 Conclusions

As a part of the work a numerical model of unmanned aircraft equipped with a 2 DoF manipulator has been developed. Tubular-beam manipulator and tests onboard the UAV were done. Based on the numerical simulation and field tests two major conclusion can be formulated.

Manipulator tests on the Aquila helicopter (with PID control system), and numerical simulations of helicopter-manipulator system (with the LQR control system) showed the impact of the manipulator configuration on the components of the vector control. Response of the control system for the manipulator movements were stronger when the manipulator was loaded by additional mass. Tests and numerical simulations showed that the helicopter is particularly sensitive to the rotation of the fully ejected manipulator, and less when manipulator is only ejected.

The range of the position which may be obtained by UAV control servo is limited. This means that the impact of the manipulator configuration can be compensated by the control system only in a limited range. Acceptable range of the manipulator position, depends on the weight carried by the tip of the manipulator, as well as the prevailing conditions (e.g. wind speed). Therefore it is not possible to define a continuous range of allowable manipulator position. It is necessary to use the current data of helicopter's state to stop the motion of the manipulator before the components of the vector control helicopter go beyond acceptable ranges. Data from the helicopter's control system must be transferred to the manipulator's control system before the range of the helicopter control servo will reach. It is possible to determine acceptable ranges of pitch and roll angle of the helicopter, above which the manipulator would be automatically stopped. Control system (PID controller based) for Aquila helicopter can be improved by changing the controller settings, depending on the current configuration of the manipulator. In addition, information from the helicopter control system should be transferred to the manipulator control system to make possible automatic stop of the manipulator's motion. Schematic diagram showing the upgraded PID control system is on Fig. 21.

Three characteristic configurations of manipulator for which the PID set controller should be selected can be assumed: folded and straight manipulator, manipulator ejected to 1 m and the manipulator ejected to 1 m and rotated by 90°. These configurations are shown in Fig. 22a. In the intermediate positions (for all other configuration of manipulator) the PID controller settings helicopter should be interpolated by an additional control system.

Comparison between the results of numerical simulations and measurements from test-flights of the helicopter show, that during motions of the manipulator, behavior of the numerical model is quite different than the of real UAV. This is due to two causes: (i) simplicity of the numerical model which does not take into account some dynamical aspects of helicopter flight and (ii) high measurement noise recorded during test-flights. However, as can be seen from Fig. 22b, changes

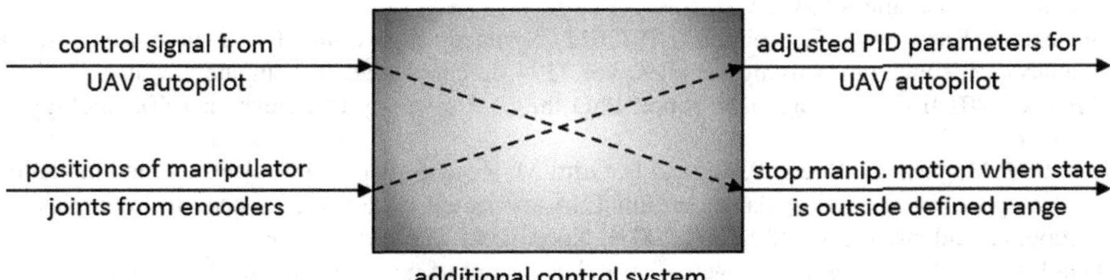

Fig. 21 Schematic diagram of an upgraded control system

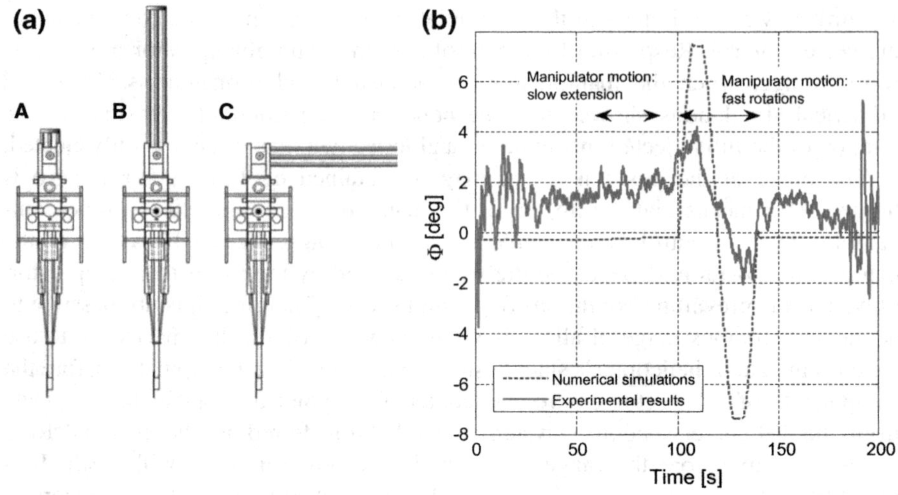

Fig. 22 **a** Three configurations of the manipulator: A—manipulator folded and straight, B—the manipulator is ejected, C—the manipulator is on 1 m ejected and rotated by 90°. **b** Comparison (right) between numerical simulations and results of test-flights: helicopter roll angle during motion of the manipulator (extension and rotation)

in the helicopter roll angle during fast rotations of the manipulator can be seen in the measurements from test-flight and in the results of numerical simulations.

Acknowledgements The paper was supported by national project no N N 509 25773 founded by polish Ministry of Science and Higher Education (MNiSW).

References

Albers A, Trautmann S, Howard T, Nguyen TA, Frietsch M, Sauter C (2010) Semiautonomous flying robot for physical interaction with environment, Robot Autom Mechatron

Bareiss D, Van den Berg J (2012) Reciprocal collision avoidance for quadrotor helicopters using LQR-obstacles. In: AAAI workshop on multi-agent pathfinding—WoMP

Bhandari S, Colgren R, Lederbogen P, Kowalchuk S (2005) Six-DoF dynamic modeling and flight testing of a UAV helicopter. In: AIAA modeling and simulation technologies conference and exhibit, San Francisco, USA

Bielecki B, Buratowski T, Śmigielski P (2012) Syntactic algorithm of two dimensional scene analysis for unmanned flying vehicles, vol 7594. Lecture notes in computer science

Chmaj G (2010) Flying diagnostic robot, PhD thesis, University of Science and Technology in Cracow

Ferretti G, Magnani G, Rocco P, Vigano L, Gritti M, Rusconi A (2004) Object-oriented modeling of a space robotic manipulator. In: 8th ESA workshop on advanced space technologies for robotics and automation 'ASTRA 2004', Noordwijk, The Netherlands

Haug E (1989) Computer aided kinematics and dynamics of mechanical systems. Volume 1: basic methods, London Allyn and Bacon

Heffley R, Mnich M (1988) Minimum-complexity helicopter simulation math model. NASA

Hald U, Hesselbaek M, Holmgaard J, Jedsen Ch, Jakobsen, S, Siegumfeldt M (2005) Autonomous helicopter—modelling and control. Aalbor University, Department of Control Engineering

K (2008) The dynamics of the satellite rendezvous and docking maneuver using nonholonomic robotic arm

Kondak K, Krieger K (2012) Closed-loop behavior of an autonomous helicopter equipped with a robotic arm for aerial manipulation tasks. Int J Adv Robot Syst

Kuciński T, Rybus T, Seweryn K, Banaszkiewicz M, Buratowski T, Chmaj G, Grygorczuk J, Uhl T (2014) Deployable manipulator technology with application for UAVs. In: Sasiadek JZ (ed) Aerospace robotics, GeoPlanet: earth and planetary sciences. Springer (submitted)

Kumar V, Michael N (2012) Opportunities and challenges with autonomous micro aerial vehicles. Int J Robot Res 31(11):1279–1291

McLean D (1991) Automatic flight control systems. Prentience Hall International, New York. ISBN-321323-34242

Mellinger D, Shomin M, Michael N, Kumar V (2010) Cooperative grasping and transport using multiple quadrotors. In: Proceedings of the international symposium on distributed autonomous robotic systems

Mellinger D, Lindsey Q, Shomin M, Kumar V (2011) Design, modeling, estimation and control for aerial grasping and manipulation. IROS, IEEE, pp 2668–2673

Menon C, Aboudan A, Cocuzza S, Bulgarelli A, Angrilli F (2005) Free-flying robot tested on parabolic flights: kinematic control. J Guidance Control Dyn 28(4)

Menon C, Busolo S, Cocuzza A, Aboudan A, Bulgarelli A, Bettanini C, Marchesi M, Angrilli F (2007) Issues and solutions for testing free-flying robots. Acta Astronaut 60

Moosavian SAA, Papadopoulos E (2007) Freeflying robots in space: an overview of dynamics modeling, planning and control. Robotica 25(5):537–547

Orsag M, Korpela C, Oh P (2012) Modeling and control of MM-UAV: mobile manipulating unmanned aerial vehicle, In: Proceedings of the international conference on unmanned aircraft systems

Padfield G (2007) Helicopter flight dynamics, AIAA Series, Blackwell Publishing Ltd, Oxford

Pounds P, Dollar A (2010) Hovering stability of helicopters with elastic constraints. In: Proceedings of the ASME dynamic systems and control conference

Pounds P, Bersak D, Dollar A (2011) Grasping from the air: hovering capture and load stability. ICRA, IEEE, pp 2491–2498

Sasiadek J (2013) Space robotics and its challenges In: Aerospace robotics, springer series on GeoPlanet: earth and planetary sciences, Springer, pp 4–12

Schaub H, Junkins J (2003) Analitical mechanics of space system. American Institute of Aeronautics and Astronautics, Reston

Seweryn K, Banaszkiewicz M, Maediger B, Rybus T, Sommer J (2011) Dynamics of space robotic arm during interactions with non-cooperative objects. In: 11th ESA workshop on advanced space technologies for robotics and automation 'ASTRA 2011', Noordwijk, The Netherlands

Seweryn K, Rybus T, Banaszkiewicz M, Grygorczuk J, Kuciński T, Buratowski T, Chmaj G, Uhl T (2012) Manipulator mounted on a satellite versus manipulator mounted on UAV helicopter—comparative study. In: AIAA guidance, navigation, and control conference, Minneapolis, USA

Shima T, Rasmussen S (2009) UAV cooperative decision and control: challenges and practical approaches

Vafa Z, Dubowsky S (1987) On the dynamics of manipulators in space using the virtual manipulator approach. ICRA 4:579–585

Vafa Z, Dubowsky S (1990) The kinematics and dynamics of space manipulators: the virtual manipulator approach. Int J Robot Res 9(4):3–21

Wood G, Kennedy D (2003) Simulating mechanical systems in simulink with SimMechanics. Technical report. The MathWorks Inc, Natick, USA

Yoshida K (2003) Engineering test satellite vii flight experiments for space robot dynamics and control: theories on laboratory test beds ten years ago, now in orbit. Int J Robot Res 22(5): 321–335

Zieliński C (2000) Implementation of control systems for autonomous robots. In: 6th international conference on control, automation, robotics and vision, Singapore

Prototype, Mathematical Model and Simulations of a Model-Making Rocket

Jacek Drewniak and Ignacy Duleba

1 Introduction

In the past, rockets were analyzed mainly by scientists working for a military industry. Recently, due to cheap, reliable and small electronic components, they are within the reach of enthusiasts of aviation. Usually, high-school students or students work in a team to launch a rocket and test its flight behavior. Their projects are recorded at national www-sites [in Poland (www.rakiety.org.pl)] and their achievements are really impressive [in 2016 a team of the Stuttgart Univ. students have launched their HEROS 3 rocket to an altitude above 30 km setting a new world record in this construction class (www.hybrid-engine-development.de)].

Modeling rockets is a great challenge for robotics which started from stationary manipulators, processed with 2D mobile robots and currently extensively exploits 3D object like drones and other flying objects. Although general principles of modeling are the same for all mechanical objects as they are rooted in laws of physics, some specific phenomena should be taken into account while analyzing each particular one.

In this paper a complete project of a model-making rocket is discussed. Although performed in an extremely small team and with severely limited founds, it contains all steps to be followed while realizing big rocketry projects. A mathematical model of a rocket was derived based on parameters taken from a prototype of the rocket. Some details of constructions are provided covering both mechanics and electronics of the rocket. The paper is organized as follows. In Sect. 2 both kinematics and

J. Drewniak · I. Duleba (✉)
Department of Cybernetics and Robotics, Wroclaw University of Science and Technology, Janiszewski St. 11/17, 50-372 Wroclaw, Poland
e-mail: ignacy.duleba@pwr.edu.pl

J. Sasiadek, *Aerospace Robotics III*, GeoPlanet: Earth and Planetary Sciences,
https://doi.org/10.1007/978-3-319-94517-0_9

dynamics of the rocket is derived based on the paper (Koruba and Osiecki 1991). In Sect. 3 a simulation of the rocket's motion is presented. The components and a hardware architecture of the prototype of the rocket is discussed in Sect. 4. Section 5 concludes the paper.

2 Mathematical Model

2.1 Kinematics

A few coordinates frames were assigned to a rocket, cf. Fig. 1. A stationary, inertial global coordinate frame is used to keep trace over a motion of the rocket. A local frame is tightly coupled with its body. The origin of the frame is set at the center of mass (CG) of the rocket with its x-axis pointing towards its nose. An aerodynamic coordinate frame (aero-frame) is used to facilitate description of equations of motion. The frame is also originated at CG but its x-axis is directed towards a current velocity of the rocket. Each of four elevons has got its own coordinate frame with the origin placed on a ring of the hull of the rocket and shifted by 90°.

Transformation matrices between introduced coordinate frames are given below: in $SE(3)$ [for definition of the transformations refer to Spong and Vidyasagar (1989)]

$$T^{loc}_{glob} = Tran(x, a_x)Tran(y, a_y)Tran(z, a_z)Rot(z, \upsilon)Rot(y, \psi)Rot(z, \phi)$$

and useful transformations between orientation parts of aforementioned frames (in SO(3))

$$\mathbf{R}^{aero}_{glob} = Rot(z, \gamma) \cdot Rot(y, \chi), \quad \mathbf{R}^{loc}_{aero} = Rot(z, \alpha) \cdot Rot(y, \beta).$$

Elevons are rotated with respect to their coordinate frame by angles $\sigma_i, i = 1, \ldots, 4$ and related to the local coordinate frame by the following transformations (in SE(3)) with fixed length parameters $x_{l_i}, y_{l_i}, z_{l_i}$

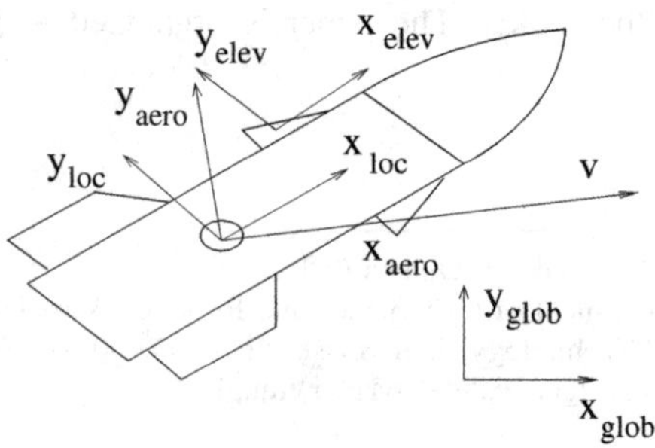

Fig. 1 Coordinate frames of the rocket

$$
\begin{aligned}
\mathbf{A}_{loc}^{l_1} &= Tran(x, x_{l_1}) Tran(z, z_{l_1}) Rot(z, \sigma_1), \\
\mathbf{A}_{loc}^{l_2} &= Tran(x, x_{l_2}) Tran(y, y_{l_2}) Rot\left(x, -\frac{\pi}{2}\right) Rot(z, \sigma_2), \\
\mathbf{A}_{loc}^{l_3} &= Tran(x, x_{l_3}) Tran(z, -z_{l_3}) Rot(x, \pi) Rot(z, \sigma_3), \\
\mathbf{A}_{loc}^{l_4} &= Tran(x, x_{l_4}) Tran(y, -y_{l_4}) Rot\left(x, -\frac{\pi}{2}\right) Rot(z, \sigma_4).
\end{aligned}
$$

2.2 Dynamics—A Linear Motion

Rocket velocity v, acting along x-axis of the aero-frame, is transformed into the global frame according to the equation

$$\dot{\mathbf{p}}_{\mathbf{g}} = \left[\dot{x}_{glob}, \quad \dot{y}_{glob}, \quad \dot{z}_{glob}\right]^T = \mathbf{R}_{glob}^{aero} \left[v \quad 0 \quad 0\right]^T. \tag{2.1}$$

Taking derivative of Eq. (2.1), and transforming the acceleration $\ddot{\mathbf{p}}_g$ into the aero-frame we get

$$\ddot{\mathbf{p}}_a = \mathbf{R}_{aero}^{glob} \ddot{\mathbf{p}}_g.$$

For close-range model-making rockets, the second Newtonian law states that the motion of CG is given in the aero-frame by

$$m\ddot{\mathbf{p}}_a = \left[m\dot{v}, \quad mv\dot{\gamma}\cos\chi, \quad -mv\dot{\chi}\right]^T = \mathbf{P}_{aero} + \mathbf{G}_{aero} + \mathbf{A}_{aero} + \mathbf{Q}_{aero}, \tag{2.2}$$

where all forces on the right hand side of Eq. (2.2) are expressed in the aero-frame. Notice that due to a low fuel mass with respect to the total mass of the rocket, it was assumed that there is no mass decrease during a flight and the half of the initial fuel mass is added to inertia parameters of the rocket. In this paper $\boldsymbol{Q} = 0$ for the linear motion. A scalar, thrust force $P(t)$ is acting along x-axis in the local frame, and a gravity force from the global frame are expressed in the aero-frame

$$\mathbf{P}_{aero} = \mathbf{R}_{aero}^{loc}\left[P(t), \quad 0, \quad 0\right]^T, \quad \mathbf{G}_{aero} = \mathbf{R}_{aero}^{loc}\left[0, \quad -mg, \quad 0\right]^T$$

The most complicated and difficult to model are aerodynamic forces $\mathbf{A}(v, \alpha, \beta)$. The most reliable estimation of their values is obtained in an aerodynamic channel based on a scaled 3D model of a rocket. In preliminary simulations, a first order linear approximation of the forces (Koruba and Osiecki 1991) were taken into account

$$\mathbf{A}(v, \alpha, \beta) = \frac{\rho}{2} v^2 - \left[C_{D_{px}} S, \quad C_{D_{py}} S_y \alpha, \quad \& C_{D_{pz}} S_z \beta\right]^T,$$

where $S = \pi d^2/4$; d—is the maximal cross-section of the rocket (caliber), S_y, S_z are lateral surfaces, ρ—is a density of the air at given flight conditions. $C_{D_{px}}, C_{D_{py}}, C_{D_{pz}}$—coefficients of aerodynamic forces depending on Mach number and angles α, β shape and areas of rudders and the shape of the rocket itself. We assumed that $C_{D_{px}} = C_{D_{py}} = C_{D_{pz}} = C_D$ and this term is a sum of a few items

$$C_D = C_{D_{fric}} + C_{D_{cone}} + C_{D_{fin}} + C_{D_{base}}. \tag{2.3}$$

For turbulent air-flow (Barrowman 1967)

$$C_{D_{fric}} = 0.032 \cdot (R_a/L)^{0.2}, \tag{2.4}$$

where R_a is a geometric parameter of the rocket (roughness) taken as 20 μm [standard value for common airplanes, Barrowman (1967)] and L is a characteristic dimension—the length of the rocket taken as 1 m. Based on Niskanen (2013) it was assumed that for the nose profile used in this project the coefficient of resistance (Haack Series) equals to C_{Dcone}=−0.05. An air-resistance due to rocket stabilizers (Hoerner 1965), for this shape and under-sonic velocities

$$C_{D_{fin}} = 0.85 \cdot \left(1 + M^{2\wedge}2/4 + M^4/40\right).$$

(M—the Mach number, i.e. a ratio of velocity of the rocket and the sound velocity) while due to low pressure following the rocket in an under-sonic flight (Fleeman 2006), the remaining coefficient equals to

$$C_{D_{base}} = 0.12 + 0.13 \cdot M^2.$$

2.3 *Dynamics—An Angular Motion*

In this paper two angular aerodynamic momenta are considered:

(1) due to lift force $\boldsymbol{Q}$ generated on controlled elevons, and projected onto the local frame,
(2) a stabilizing momentum $\boldsymbol{\Omega}$ generated on a lateral surface of the rocket, mainly on stabilizers.

The momentum $\boldsymbol{Q}$ is a sum of vector products of a lift force generated on each elevon with an appropriate lever arm when both the data are expressed in the local frame.

The lift force $C_z(\eta, v_{elev})$ depends on an angle of attack of elevon. To compute the force, angle of attack α and side-slip angle β are used to project air force into the frame of an elevon

$$\left[v_{x_{elev}}, \quad v_{y_{elev}}, \quad v_{z_{elev}} \right]^T = \mathbf{R}_{elev}^{loc} \cdot \mathbf{R}_{loc}^{aero} \cdot [v, \quad 0, \quad 0]^T.$$

Only the velocity v_{elev} on the xy plane and the angle η are of interest:

$$v_{elev} = \sqrt{v_{x_{elev}}^2 + v_{y_{elev}}^2}, \qquad \eta = a\tan^2\left(v_{x_{elev}}, v_{y_{elev}}\right).$$

Based on the same principle, a momentum is generated on stabilizers. As stabilizers are fixed objects in the local frame, the momenta can be expressed directly in the local frame

$$\Omega_y(v, \alpha, \beta) = [0, \quad \Omega_y(v, \alpha, \beta), \quad \Omega_z(v, \alpha, \beta)]^T. \tag{2.5}$$

Functions in Eq. (2.5) are usually derived via simulations but similarly to A, their simplified versions that depend on flight parameters can be derived (Koruba and Osiecki 1991)

$$\Omega_y(v, \alpha, \beta) = -C_{Dx_{fin}} v^2 \beta, \quad \Omega_y(v, \alpha, \beta) = -C_{Dy_{fin}} v^2 \alpha.$$

Assuming symmetry, the coefficient $C_{Dx_{fin}} = C_{Dy_{fin}} = C_{D_{fin}}$ depends on geometric parameters and the Mach number M. Finally, all momenta form the resulting momentum

$$\mathbf{M} = \sum_{i=1}^{4} \mathbf{A}_{loc}^{elev_i} C_{Zl i}(v_i, \eta_i) + \mathbf{\Omega}(v, \alpha, \beta) = Q + \mathbf{\Omega}.$$

The generalized orientation coordinates of the rocket in the global frame are given by RPY angles $\mathbf{q} = (\phi, \psi, \vartheta)^T$ and an angular velocity ω of the local frame is introduced. The Lagrange equation of an angular motion is based on the kinetic energy preservation law $E_k = 1/2 \cdot \omega^T I \omega$ with an inertia matrix $\boldsymbol{I}$ expressed in the local frame:

$$I\dot{\omega} = M \Rightarrow \dot{\omega} = I^{-1} M. \tag{2.6}$$

The angular velocity ω in the body frame (Tchon et al. 2000) is retrieved from well-known skew-symmetric matrix $\boldsymbol{S}(\omega)$ defined by

$$S(\omega) = \mathbf{R}_{glob}^{loc\ T} \cdot \dot{\mathbf{R}}_{glob}^{loc} \quad S(\omega) \Rightarrow \omega = \begin{bmatrix} 1 & 0 & -\sin\psi \\ 0 & \cos\phi & \cos\psi\sin\phi \\ 0 & -\sin\phi & \cos\psi\cos\phi \end{bmatrix} \begin{bmatrix} \dot{\phi} \\ \dot{\psi} \\ \dot{\vartheta} \end{bmatrix} = A_\omega \dot{\mathbf{q}}. \tag{2.7}$$

Table 1 Forces and drag coefficients for some flight parameters: C_D from Eq. (1.3) C_m—an elevon lift force coefficient

Flight parameters		Forces and coefficients			
α (°)	v (m/s)	F_{Cm} (N)	$C_m(\alpha)$ (–)	F_{CD} (N)	C_D (–)
0	30	0	NA	1.74	0.0019
	100			20.07	0.0020
	200			61.07	0.0038
4	30	−0.28	0.05	2.08	0.0023
	100	−3.15		20.64	0.0020
	200	−8.68		57.67	0.0014
10	30	−0.71	0.05	2.16	0.0024
	100	−7.81		22.81	0.0023
	200	−28.05		72.22	0.0018
30	100	−16.95	0.038	36.27	0.0036
−4	30	0.27	0.05	1.79	0.0022

Taking derivative of Eq. (2.7) and substituting it into Eq. (2.6) we get

$$\ddot{\mathbf{q}} = A_{\omega}^{-1} \cdot (I^{-1} \cdot M - \dot{A}_{\omega} \cdot \dot{\mathbf{q}}).$$

Notice that due to parameterization of $SO(3)$ with RPY angles det $\boldsymbol{A}_{\omega} = \cos\psi$ and singular configurations (known as a gimbal lock) may appear for $\psi = \pm\pi/2$. For this case, it is proposed to use a technique developed in Duleba (1996). It prompts to change a parameterization (for example from RPY to Euler angles or vice versa) when a currently used parameterization becomes ill-conditioned. Singularities for both parameterizations do not appear close to each other in $SO(3)$. Consequently, a minimally dimensional parameterization of $SO(3)$ is preserved and singularities are avoided.

It can be noticed that aerodynamic forces depend strongly on some coefficients and, in many cases, are based on simplifying assumptions. In simulations, it was observed an overestimation of the drag coefficient for stabilizers. Therefore the computational fluid dynamics (CFD) for the rocket was run. A tetragonal mesh, implemented in Salome API within Python script, covered its 3D model. Then, the mesh was loaded into the openFOAM project and computed by steady-state solver for incompressible, turbulent flow—simpleFoam. The calculations were made for different inclinations of elevons and a fluid speed. Table 1 shows that drag coefficients are significantly smaller than prompted by the analytical model, even with a non-aerodynamic shape of stabilizers. Results presented in the table also confirm that elevated elevons generate pitching momenta—confirming analytical calculations and proving their ability to influence a rocket's motion.

3 Simulations

A mathematical model of the rocket was implemented in Mathematica package, well suited for symbolic and numeric computations. No particular optimization of the code was performed. Exemplary simulation plots, depicted in Fig. 2, cover the case when the rocket set vertically was launched with the speed of 10 m/s. One pair of elevons—perpendicular to the XY plane—were fixed to 8° on the interval (0–5) s. The engine thrust was set to 30 N on the interval (0–2) s and linearly decreased to 10 s, afterwards.

In Fig. 2a, it can be seen that elevons gradually slope the trajectory in the direction of the x-axis. The torques due to elevons and opposite torques caused by stabilizers are depicted in Fig. 2g, h. In the figure, it can be observed that in 5 s the flight is stabilized as the resultant force is close to zero, and the rocket flights with a constant angle of attack with the torques canceled. Further simulations show that the angle of attack of the rocket is proportional to the angle of attack of elevons. The apogee of ascendance is reached near 20 s where the rocket loses a vertical speed. Its nose turns down and begins to sway as depicted in Fig. 2b, d.

4 Prototype

A standard model-making rockets are designed to fly and return safely using stabilizers and a parachute (www.rakiety.org.pl). The prototype of the actively controlled rocket was made with lightweight materials to satisfy NAAR Rocket Safety Code. As the rocket is powered with a factory made engine up to 160 Ns, it can be used without any special license. The fully assembled rocket is visualized in Fig. 3.

A cylindrical part of the rocket was made with layers of a special rigid cartoon hardened with a glue. The poliactide nose has a shape of Haack series. Four trapezoidal stabilizers made with balsa were screwed and glued to the bottom part of the body, nearby the engine socket. The prototype is powered by a solid fuel rocket engine. The Control Module, designed from scratch and made by the first author, Drewniak (2015), is placed close to the nose cone. It is composed of four elevons, a controller, a battery and a cover. Elevons generate a lift force when appropriately oriented. The forces applied at some distance from the center of mass can generate torques in all directions.

The main component of the module is a controller, divided into two sub-parts. The upper part, cf. Fig. 4, is responsible for calculating parameters of a flight, trajectory and appropriate controls for elevons. The bottom part, Fig. 5, is responsible for setting required orientations of elevons. All functions of the controller are depicted in Fig. 6. A memory module was designed to store parameters of the flight as well as commanded controls. Those data can be used in off-line mode to verify a mathematical model implemented and to estimate currently unmodeled phenomena like a wind speed (to improve the model in future flights). Vital parts of the controller were located far from the power section to avoid its noise-making influence.

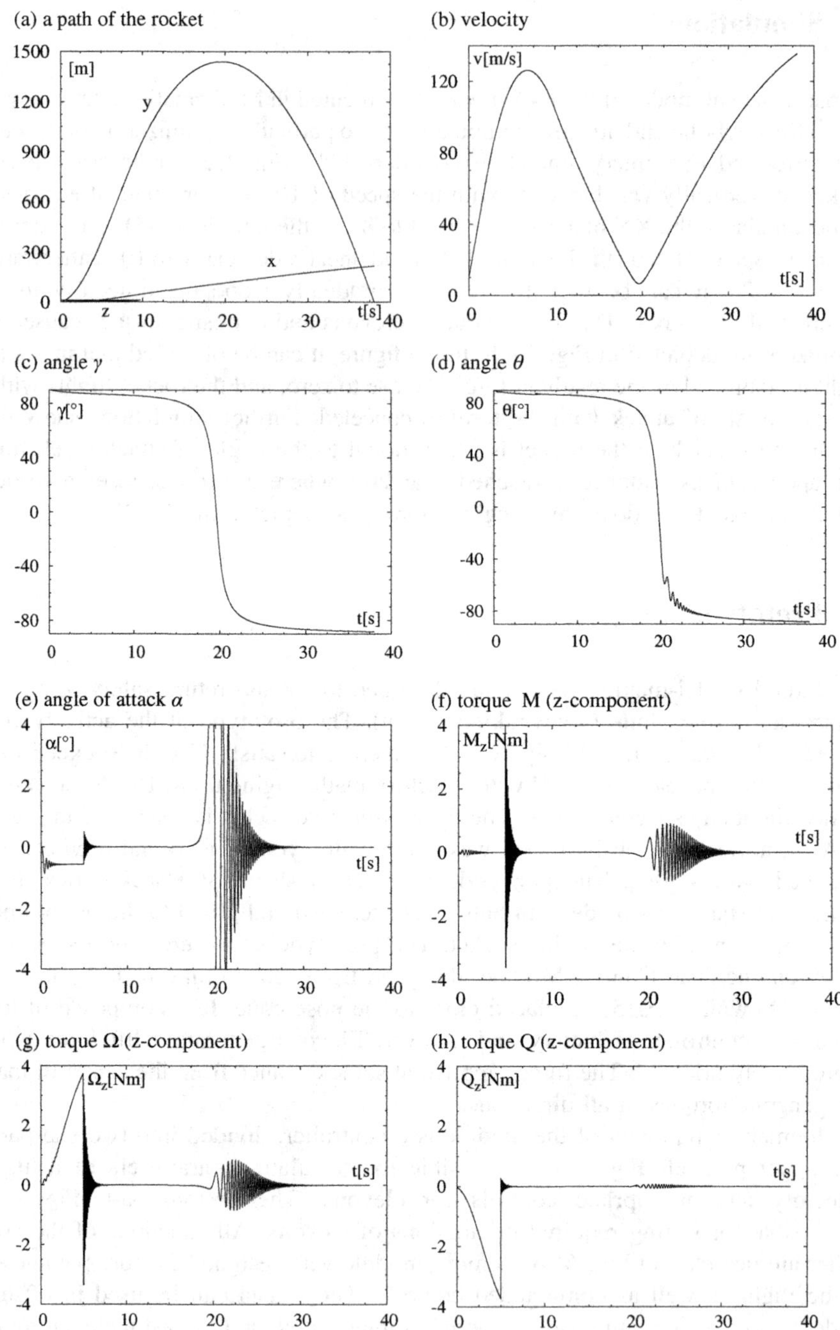

Fig. 2 Simulation results

Fig. 3 A fully assembled rocket

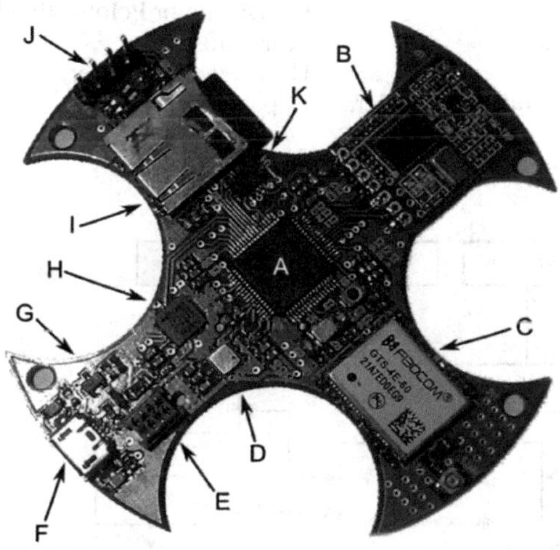

A - STM32F103RET6 microcontroller,
B - HMR-TRP-868 radio,
C - GPS GTS-4E,
D - BMP180 barometer,
E - micro-match socket,
F - USB socket,
G - extra power source,
H - CP2102 converter,
I - micro-sd slot,
J - SWD socket,
K - BMX055 9dof inertia sensor.

Fig. 4 Upper part of the controller

5 Conclusions

In this paper an almost complete project of a model-making rocket was presented (still it waits for the launching). It was developed in a relatively short amount of time but with a lot of work done at many stages of the project. Such projects are

A - CP2102 converter,
B - USB socket,
C - TB6612 h-bridge,
D - SWD socket,
E - STM32F103C8T6 microcontroller,
F -I/O section,
G - DC motor Pololu HP with 300:1 gear ratio,
H - magnetic encoder,
I - power section.

Fig. 5 Bottom part of the controller

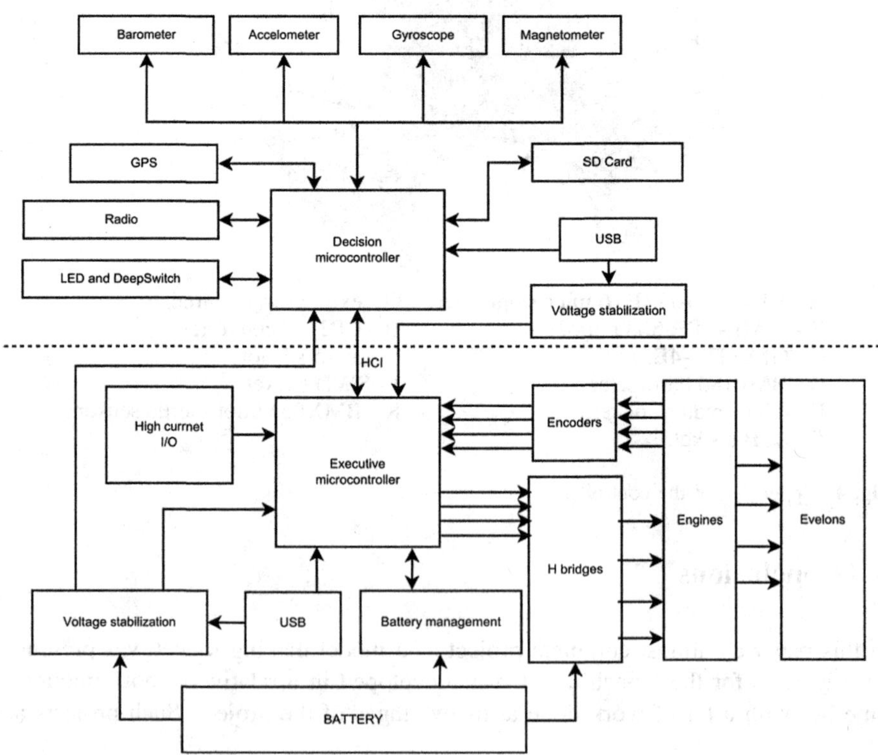

Fig. 6 A functional scheme of the controller

especially difficult to realize in small teams as they require knowledge and skills from many different domains of science: mechanics, electronics, mathematics, computer science. Many tools, supporting software programs should be applied and some experts in particular domains should be consulted to successfully finish the project. Finally, some spectacular effects can be obtained as well as teaching effects fulfilled.

References

Barrowman JS (1967) The practical calculation of the aerodynamic characteristics of slender finned vehicles. MSc thesis, The Catholic University of America

Drewniak J (2015) Modeling and visualization of a model-making rocket. MSc thesis, Wroclaw University of Technology (in Polish)

Duleba I (1996) On avoiding singularities of representation. In: Works of IX symposium on simulation of dynamical processes, Zakopane, pp 345–350 (in Polish)

Fleeman E (2006) Tactical missile design, 2nd edn. AIAA Education Series

Hoerner SF (1965) Fluid-dynamic drag. Bricktown, New Jersey

Koruba Z, Osiecki JW (1991) Construction, dynamics and navigation of close-range rockets. Kielce University of Technology Publishing House (in Polish)

Niskanen S (2013) OpenRocket technical documentation. MSc thesis

Spong MW, Vidyasagar M (1989) Robot dynamics and control. Wiley, New York

Tchon K, Mazur A, Duleba I, Hossa R, Muszyński R (2000) Manipulators and mobile robots: models, motion planning, control. EXIT Publishing House, Warsaw (in Polish)

www.rakiety.org.pl

www.hybrid-engine-development.de

especially [illegible] in [illegible] such as [illegible] and [illegible] from different domains of science: mechanics, electronics, thermodynamics, computer science. Many [illegible] supporting software [illegible] should be applied and [illegible] in practical [illegible] should be [illegible] [illegible] as well as [illegible] [illegible]

References

[illegible] (19[illegible]) [illegible] calculations of the [illegible] characteristics [illegible] [illegible] University of [illegible]

[illegible] (20[illegible]) Modeling and simulation of a [illegible] [illegible] University of Technology [illegible]

[illegible] (20[illegible]) [illegible] representation. In: [illegible] [illegible] and [illegible]

[illegible] (2000) [illegible] 2nd edn. [illegible]

[illegible]

[illegible]

[illegible]

[illegible]

[illegible]

[illegible]

[illegible]

[illegible]

Space Mining Challenges: Expertise of the Polish Entities and International Perspective on Future Exploration Missions

Marta E. Wachowicz, Patrycja Frąk, Adam Węgłowski, Zbigniew Burdzy, Roger Bachtin, Joanna Bankiewicz and Wojciech Gołąbek

1 Introduction

Earth's population is growing. Simultaneously, as a planet, Earth has a finite amount of resources, and it is going to run out of them. It is reasonable to start planning and preparing for this inevitability as soon as possible. Drawing assets from outer space is a promising alternative way. Mining the asteroids and other small celestial bodies is not just the matter of science fiction or writers' rich imagination. The title in The Guardian (Tuesday, December 6, 2016) *Galactic gold rush: the tech companies aiming to make space mining a reality*, perfectly reflects the trend among the visionary companies, and it seems that the era of space mining is already underway. Space mining can be treated as a combination of technological challenges and human ambition indicators in space exploration beyond Earth at the edge of the Solar System. The issue of space mining is a consequence of the development of new deep space exploration ideas; it raises the issues concerning humankind rights to colonize other planets and global environmental protection. The main benefits of such activities involve extraction and processing of space resources into useful products. The ability to make new fuel (propellants) or infrastructure can reduce the cost, risk, and mass of robotic and later human participants of exploration beyond Earth while providing capabilities which will enable the commercial development of space. Governmental authorities and private companies are already working on asteroid mining projects. Luxembourg recently established a special fund for space mining projects; the United States approved the Commercial Space Launch Competitiveness Act as a binding national law, and private American companies such as Planetary Resources and Deep Space Industries are preparing the technologies to extract resources in a space environ-

M. E. Wachowicz (✉) · P. Frąk · A. Węgłowski · Z. Burdzy · R. Bachtin · J. Bankiewicz
W. Gołąbek
Polish Space Agency, Powsinska St. 69/71, PL02-903 Warsaw, Poland
e-mail: marta.wachowicz@polsa.gov.pl

J. Sasiadek, *Aerospace Robotics III*, GeoPlanet: Earth and Planetary Sciences,
https://doi.org/10.1007/978-3-319-94517-0_10

ment. One of the major concerns with space mining is, of course, landing on the surface of celestial bodies such as asteroids and the development of technologies which are necessary to identify, extract, and process minerals, water, and other valuable assets.

The paper is structured through seven sections as follows. The next section describes the technological problems, obstacles, and progress of R&D works devoted to robotic systems enabling autonomous movement on the surface of the planet or asteroid, manipulators, and microgravity-driven instruments for probes sampling. To give the broad context of the issue, a wide range of examples is presented. The next section is focused on economic value of new space-based products. The fourth section presents the space heritage related to planetary exploration and presents European, American, Chinese, Japanese, and Russian achievements and their national perspective towards space mining. The fifth section analyzes Polish experience in exploration mission and entities of the Polish market which are developing space mining oriented R&D works. Particular attention is paid to the potential of Polish space industry in the field of robotics, subsurface research, drilling mechanisms, and devices. The sixth section concentrates on intellectual property issues and legal aspects. The paper shows the importance and applicability of the provisions of the international treaties with respect to the exploitation of extraterrestrial resources.

2 Technological Challenges of Space Mining

2.1 Reasons for Space Mining

Space mining or In-Site Resource Utilization (ISRU) is the area of activity related to collection of materials derived from the celestial bodies of the Solar System and the extraction activities performed on the surface of extraterrestrial objects. Space mining is a very forward-looking and future oriented field of space activities. Moreover, it is a crucial research and development discipline of space activities. The whole process of collecting, processing, storing, and using materials encountered by space exploration, jointly with the utilization of native resources to produce new products or new materials, is described by the term ISRU (Sacksteder and Sanders 2007). According to Sanders, ISRU involves any hardware or operation that utilize in situ resources to create products or services for space exploration (Sanders et al. 2017).

First, it is an impulse for the development of space technology in the robotic systems or drilling devices, enabling to take samples directly from extraterrestrial bodies, exploration of extraterrestrial small bodies of the Solar System, such as asteroids or moons. Second, space mining can be one of mankind's ideas to secure electrical energy for Earth's total population. Another argument for developing space mining and technological areas related to this issue can be a need to build a base or sustainable infrastructure on other planets or asteroids. Developing bases or habitats may facilitate the acquisition of distant planets (e.g., the Moon base or Moon village can serve as a transit station on the way to Mars), or prepare to create alternative housing for people due to overpopulation of Earth as a planet in the distant future. Considering the long-lasting process of preparing exploration missions and the Polish space heritage in subsurface instruments and technological demonstrators concerning the drilling processes, the development of R&D works in several key areas is necessary.

To understand and recognize the complexity and difficulties of the whole space mining activities, the following constrains should be mentioned: astrodynamics (destination of the mission, selection of the planetary body); environmental conditions; payload mass and power resources; service life and maintenance of dedicated devices; and lastly, mission scenarios with transport and orbit modifications. The development of mining and the process of site preparation requires first reaching the near-Earth object, then motion control and emplacement of the restraint system, and constructing the operational and processing systems (Sanders et al. 2017; Crites et al. 2012). Second, the process of extraction should consist at least of mining, processing, and transport, which changes simultaneously the main body motion and orbit modification (Blair and Gertsch 2010).

Asteroid or lunar resources can provide many critical elements such as fuel for space vehicles or materials for construction of habitats or bases. It is obvious that mining techniques and terrestrial devices will not work in space, although terrestrial concepts provide a starting point for space-oriented R&D. There is a fundamental similarity between processes related to terrestrial and extraterrestrial mining. Consequently, modified terrestrial approaches can be adopted. For both space and terrestrial mining, the first step is prospecting, globally and locally, to find the resources in the geological context, then mapping, and finally, mining and resource processing. As with terrestrial mining, subscale feasibility as well as pilot operations are performed to verify extraction of the resources (Sanders 2015).

2.2 *Obstacles in Space Mining Development and Challenges in Space-Oriented R&D*

The following problems should be faced in all phases of mining performed on asteroids or other planetary bodies:

- environmental constraints—low gravity, dust, vacuum, temperature gradient, radiation—have a strong influence at effectiveness of devices, mining methods, and mission operational difficulties (Zacny et al. 2008; Hoffman et al. 2016);
- remote operation and automatization processes is necessary;
- lack of geotechnical engineering data for processing sites;
- engineering properties of the lunar regolith including density, compressibility, permeability, shear strength, cohesion, diffusivity, bearing capacity and

trafficability, particle size distribution, particle shapes, bulk density, porosity, slope stability, as well as geochemical features (Chamberlain et al. 1993)—discussion about regolith research can be found in literature (Humphries 2013; Colwell et al. 2007; Carrier 2003; Carrier et al. 1991; Skonieczny et al. 2014; Carrier and Mitchell 1989).
- all steps such as mission preparation, specific devices building—site preparation, excavations, mining, waste disposal, finally transporting, and coming back to the Earth should be considered (Chamberlain et al. 1993).

The variety of ideas regarding how to drill and take the samples in extraterrestrial environments and the overview of space instruments is described in the review book *Drilling in Extreme Environments: Penetration and Sampling on Earth and other Planets* (Bar-Cohen and Zacny 2009). The most common and already used tools (or being developed for the exploration mission) are surface drills, grinders and rock abrasion tools, scoops, all types of moles, ultrasonic and percussive actuated drills, hammering mechanisms, and regolith penetrometer (Bar-Cohen and Zacny 2009). Polish space heritage is strictly connected with designing, testing, and preparing for the space missions hammering devices and the development of the whole family of penetrometers (see Sect. 5.1). In addition, the Rosetta mission can be shown as an example of step forward for the prospects of space mining. Rosetta's probe–Philae, the first-vehicle to dock on the surface of a comet, was equipped with MUPUS (MUlti PUrpose Sensor for surface and subsurface science)—a hammering device made by the Polish engineers (see Sect. 5.1). After the success of the Rosetta mission, the perspective for future robotic missions seems to be more real and substantial.

There are many planetary bodies that are being considered for in situ exploration missions, including Mars, and the Moon, but other satellites of Solar System planets are also being considered. Specifically, during the current National Aeronautics and Space Administration (NASA) decadal mission planning and ESA effort, Mars has been identified as a significant scientific target for surface in situ sampling missions (ESA/PB-HME-17 2016). To support future missions, these sample handling technologies must be developed to meet a broad range of potential requirements including a variety of rock or subsurface materials, the presence of dust, limited resources, and rigorous sample preservation and the general problem of autonomous operation (Zacny et al. 2008).

Space mining implementation challenges can be summarized, dividing into 3 groups of problems which must be solved. The first involves the problem of existing of proper resources in a right amount and proper form, distribution and contaminants of the resources/ore, and what is strictly connected with the above-mentioned issue—feasibility to collect, extract, and process the potential resources (Skonieczny 2013; Skonieczny et al. 2013, 2014).

The second group of obstacles concerns the necessity to optimize the process of maximizing of the performance simultaneously with the minimisation of mass of the devices and infrastructure used to perform the operation, especially for long duration, autonomous mission, and regarding the reliability of the system, including unknown environmental conditions, and operation in extreme environment (radiation, temperature, dust, pressure, especially low gravity with possible domination of friction, and cohesion forces in microgravity) (Zacny et al. 2008).

The third group of difficulties is devoted to logistical and transportation aspects on Earth and having constant transport services, what requires the development of business model and immediate and long-term return on space investments (Sanders et al. 2017). Overcoming these challenges requires a multidisciplinary, integrated and long-term approach (Sanders et al. 2017). Multinational cooperation is also necessary.

Space-oriented R&D works are crucial. As the most relevant, the following fields of technology have been estimated:

- space robotics, especially robotic systems enabling autonomous movement on the surface of planet or asteroid, manipulators, control systems, and manoeuvrability on a surface of planet;
- drilling systems, sampling and transportation systems, penetrometers, microgravity-driven instruments for sampling, analysis and transport of raw materials/regolith;
- surface and subsurface studies of planets and small bodies of the Solar System, including analysis of loose and regolith materials, planetary analogues, planetary subsurface modelling, planetary geology, and geochemistry;
- planetology, including methods of analysis and diagnosis of planetary atmospheres;
- habitats and the formation of structures from regolith materials, including 3D printing.

3 Economic Aspects

Many terrestrial resources are running out, and there is a risk that high technologies will not be developed because of the lack of raw materials and competition for remaining resources. Environmental damage caused by extraction of poorer and more problematic deposits can cause regional or global conflicts. Utilization of asteroid resources may provide a partial solution to the problem (Ross 2001).

Technology development each year has affected humanity in many different ways. One of the consequences is overpopulation, where the number of existing human beings exceeds the carrying capacity of our planet. Constantly improved medical treatment or reduced mortality rate are only a few of the causes of overpopulation. Increasing the lifespan and the growth of the population causes new risks with that process that requires tackling the global challenges to solve new fundamental problems of the human community.

The most important issues are as follows:

- searching of alternative energy sources;
- shortage of critical mineral stocks on Earth;

- necessity of building habitats on celestial bodies;
- reduction of population density on Earth and plans of colonization of other planets and bodies of the Solar System.

Determining what could be mined, three kinds of resources from an economic perspective are precious: valuable minerals (so called Rare Earth Elements), new propellants (water, petrochemicals), and construction materials for habitats (regolith, soil).

3.1 Rare Earth Elements

Rare Earth Elements (REE) comprise a group of 15 Lanthanides, as well as Scandium and Ytterbium; light REE (LREE) comprise Lanthanum, Cerium, Praseodymium, Neodymium, Promethium; and heavy REE comprise Samarium, Europium, Holmium, Thulium, Lutetium, Scandium and Ytterbium (Ross 2001). These metals are characterized by similar physical and chemical properties and are essential in the production of fiber optics, lasers, catalysts, X-ray photomultipliers, and solar cells. REE are used for many commercial applications including new energy technologies, electronic devices, automobiles, and national security applications. The diverse properties of the REE have led to an increasing variety of applications, ranging from lasers, magnets, batteries, magnetic refrigeration, high-temperature superconductivity, and safe storage and transport of hydrogen (Haxel et al. 2002). Contrary to their name, they are not rare elements in the Earth's crust, but their considerable dispersion is crucial. REE can occur with other elements such as gold, uranium, phosphates, copper, and iron and have often been produced as a byproduct. REE are obtained from two minerals: bastnaesite (95%) and monazite (Rare Earth Elements Profile 2010). The process of acquiring them is complex because of their low concentration, depending on the chemical composition of the ore and depending on the available processing methods. LREE are more abundant and concentrated and usually make up about 80–99% of a total deposit. REE form a chemically coherent group, and their versatility and specificity has given them a level of economic and technological importance considerably greater than might be expected from their relative obscurity (Rare Earth Elements Profile 2010). Several REE are essential constituents of both petroleum fluid cracking catalysts and automotive pollution-control catalytic converters. Use of REE magnets reduces the weight of automobiles. In many applications, REE are also advantageous due to their relatively low toxicity.

The supply chain for REE generally consists of mining, separation, refining, alloying, and manufacturing. Bastnaesite deposits in the United States and China account for the largest concentrations of REE, while monazite deposits in Australia, South Africa, China, Brazil, Malaysia, and India account for the second largest concentrations of REE (Haxel et al. 2002). Bastnaesite occurs as a primary mineral, while monazite is found in primary deposits of other ores and is typically recovered as a byproduct. Over 90% of the world's economically recoverable REE are found in primary mineral deposits (i.e., in bastnaesite ores). In 2016, excess of global supply caused the decline of prices for many rare-earth compounds and metals. China has continued to dominate the global supply of rare-earth compounds and metals (Gambogi 2017). In China, the rare-earth mining production quota for 2016 was set at 105,000 ton, remaining unchanged from 2015. China's rare-earth elements industry continued its consolidation into six major industrial entities. Through September 2016, China had exported 35,200 ton of rare-earth materials, recording a 50% increase compared with exports for the same period in 2015 (Gambogi 2017). The United States was self-sufficient in the critical materials but over the past decade has become dependent upon imports (Haxel et al. 2002).

According to the United States Geological Survey (USGS), China holds 50% of the world's reserves (55 million metric ton out of 110 million ton) and the United States holds about 13%; Africa and Canada also have significant REE potential (Haxel et al. 2002). REE reserves are also found in Australia, Russia, Brazil, India, and South Africa (Humphries 2013). Currently, the European industry is 100% dependent on imported REE. There is no active rare-earth excavation in Europe, but work is being done on the use of sources in Europe. The European Commission finances project European Rare Earths Competency Network (ERECON) aimed at increasing the independence of the European Community countries in the supply of REE materials. The European Commission has created lists of critical materials, which delivery is related to high risk of disruption or suspension of supplies.

The ERECON report presents the historical and future global anticipated demand for rare-earth metals. According to the United States Geological Survey (USGS), world output in 2014, 2013, and 2012 amounted to 123,000, 110,000, and 110,000 ton, respectively. The ERECON report predicts that REE demand will grow by more than 20% in 2017 and 50% by 2020 compared to 2014 (ERECON 2014). This is related to environmental requirements and development of hybrid cars, wind turbines, and high-performance lighting systems that require REE. The main European REE exploration entities identified by the ERECON report are Solvay Rare Earth (France), Silmet (Estonia), Vacuumschmelze (Germany), NeoRem (Germany–Finland), Magneti Ljubljana (Slovenia), Treibacher Industrie (Austria), and Less Common Metals (United Kingdom). Most of the companies listed above have years of experience in REE materials and have developed the technologies necessary for extracting and processing materials. In Poland, REE occur in the Sudetes, in the area of Szklarska Poręba, and Markocice, in South-Eastern Poland in the Tajna Massif. The deposits found in Poland are mainly related to monazite, xenotime, apatite, and carbonatites. REE are also associated with phosphogypsum waste. In Poland, apatite phosphogypsum from the heap of former ZCh "Wizów" can be considered as the potential source of REE. Smaller phosphogypsum sites, which are potential sources of REE, are located in Gdańsk and Police (Całus Moszko and Białecka 2012). Całus Moszko and Białecka showed the greater concentration of lanthanides in the ash formed from the combustion of hard coal (concentration of several to several dozen times higher than its content in hard coal), so it may become a potential new source for acquiring REE (Całus Moszko and Białecka 2012).

REE, as well as critical minerals and materials, are key components of innovative economy. There is a need to broaden the REE supply chain to secure the independence and additional sources of materials—space mining could play an important role in the whole process in the future. Advanced technologies for security, telecommunications, clean energy issues, and medical devices strongly rely on raw materials from mines on Earth and on small bodies in the Solar System. The potential risk of supply disruption of critical resources has a strong impact on fragility of commodity markets, which is considerably more significant in the context of the additional space exploration risks.

3.2 New Propellants

The necessity to secure adequate amounts of fuel and to transport large amounts of water to orbit and to provide appropriate biological parameters for manned missions or space stations is defined as the fundamental problem of long-term space missions. Rocket fuel storage provides a limited supply of propellant for space vessels, so the production of rocket fuel on asteroids would allow missions to probe deeper into space. This possible solution could reduce the cost and difficulty of that activities significantly, allowing more efficient exploration of deep space (Meyers 2015). Water is a key factor in sustaining the life process, but it can also be used as a propellant. Electrolyzed water molecules produce oxygen for breathing and sustaining combustion, and hydrogen in the gaseous state can be transformed into a liquid state and may be used as a propellant. Research is underway on the Sabatier reaction where methane and water are obtained from carbon dioxide and hydrogen. The main idea behind obtaining new fuels for space activities is the secondary use of hydrocarbons. Among chemical concepts, research has been greatly influenced by the work on the Fischer-Tropsch synthesis (Davis and Occelli 2010), a catalytic chemical reaction of the formation of hydrocarbons from a mixture of carbon monoxide and hydrogen, or synthesis gas has been highly influential.

Mission Lunar Reconnaissance Orbiter and Lunar Crater Observation and Sensing Satellite, both being NASA's missions, have already found substantial amounts of ice in permanently shadowed craters on the Moon. Ice is located in cold and dark regions of the Moon, where no sunlight is available to power roving vehicles, so there is a plan to install big mirrors on the craters' rims to illuminate solar panels in the permanently shadowed regions. Depending the location of the optimal ice reserves, there will be a need to build several small robotic moon bases for mining ice, manufacturing liquid propellant, and transferring it the spacecraft. Researching the technologies and operations associated with drilling into icy deposits and extracting water from those deposits is still going on, NASA projects are worth mentioning (Hoffman et al. 2016).

The most valuable resource contained on the Moon is Helium-3, an isotope scarcely present on Earth. Helium-3 combined with for example, deuterium, can be used as fuel in fusion power reactors. The value of usage of that element is that it can generate nuclear power through a process of nuclear fusion which does not produce toxic waste. R&D works related to the estimation of helium-3 spatial distribution on the lunar regolith layer are complex (Jin and Fa 2007).

3.3 Construction of Habitats on Celestial Bodies for Exploration of Other Planets

National space agencies plan to send a man to the Moon at the end of the next decade. Manned missions are part of ESA's projects planned after 2030 as a part of the third period of European Exploration Envelope Program (E3P) (ESA/PB-HME-17 2016). The reason for that much interest in the Moon is the plan to use the raw materials there for future long-term space missions and its economic benefits. The presence of frozen volatile compounds and ice water of unknown origin has been confirmed on the Moon's surface (MOON 2020-2030 2016).

Planned missions to study poles and shadowed parts of the Moon aim to provide more detailed information. ESA plans to work with national space agencies to build an international space base on Earth's natural satellite (Neal 2015). Private actors have also submitted plans for Moon exploration over the next decade, including Airbus and Boeing (lunar testing platforms, lift systems) or Google (lander—Xprize) (Barton 2015). ESA is currently conducting research into the space environment and development of the technologies necessary for human beings to live on the Moon. Building a base on the Moon—the concept of Moon Village—is considered as a milestone for future manned missions to Mars. Currently, ESA is also working on the Human Habitat Inflatable Module, which is designed to allow people to stay on the Moon's surface. Work on temperature maintenance control systems is in progress, and projects are designed to use ISS and Orion ESM systems (Gaubert 2015). Presently, ESA is working on technologies expected for utilization in the Moon Village project and others future habitats on celestial bodies, which most are going to be implemented in the context of *Exploration Preparation, Research, and Technology Programme* (ExPeRT program) (ESA/IPC-69 2017).

A rarely mentioned issue is the need to determine and specify the economic demand model and business model of space mining activities. The most relevant question is how the conditions for REE, platinum group metal, and water mining can be profitable after returning these to Earth. According to Cohen (Cohen 2013), the cost of REE mining from space objects, for example, asteroids, is very expensive, which makes mining unprofitable in the near term. Calculations for platinum metals are more profitable, and water mining seems to be the most beneficial technologies that can establish a sustainable new-space propellants business.

4 Overview of Space Exploration Missions Related to Space Mining Challenges

4.1 European Space Mining Related Activities

For decades, the European Space Agency has been involved in deep space exploration and has been successfully developing new technologies and devices concerning robotic and human missions beyond the Low Earth Orbit (LEO). The main goal has been ensuring the European leadership in high technologies and innovative solutions regarding space on the global market. The key exploration missions are listed as follows:

- GIOTTO—The first ESA deep space mission, which launched in 1985. In 1986, Giotto passed the comet Halley nucleus as close as possible. The main goal of the mission was to help solve the mysteries surrounding Halley. Halley comet, which was one of the over 1000 then-known comets, but it was unique being young and active, as well as having a well-defined path. Giotto took the first pictures of the comet Haley shape, what gave data about composition and structure of the comet. To change its orbit, Giotto used the Earth gravitational assist. This maneuver was performed for the first time during a space mission. After the flyby, many instruments were damaged, and mission was directed to the comet Grigg-Skjellerup. In 1992, the mission ended (Wilson 2005).
- CASSINI–HUYGENS—The first ESA exploration mission, during which, one of the objectives was to land on the surface of Titan, one of Saturn's natural satellites. The Cassini–Huygens mission was implemented in cooperation with NASA and Italian Space Agency. The mission was launched in 1997, and in 2005, it reached the goal. Huygens probe disconnected from Cassini orbiter and was dropped on the surface of Titan. The probe measured Titan's atmosphere and the surface around the landing place. Huygens sent also many pictures of the surface of Titan. After this mission ended, Cassini studied Saturn and his rings until 2008 (Wilson 2005).
- SMART-1—Smart-1 was planned as a test for new technologies needed for deep space missions such as Bepi Colombo. The mission was launched in 2003 and was intended to reach the Moon orbit and take pictures of its surface. The mission tested future communication techniques for spacecraft and techniques to achieve autonomous spacecraft navigation in deep space exploration missions. The Smart-1 made high resolution pictures of the Moon and performed precise mapping of its surface. During the mission, the geological structure of the Moon surface was also studied. The probe analyzed the lunar regolith in the context of the occurrence of water, magnesium, aluminum, and silicon. The mission ended in 2006 (ESA Webpage 1).
- MARS EXPRESS—The first ESA mission to Mars was launched in 2003. Looking for water and mineral elements was the main aim of the mission. After losing the lander, the mission was continued. Mars Express studied the structure

of the planet surface and the atmosphere and chemical composition of the natural Mars satellites during the flyby near Phobos and Deimos. The mission gave a large amount of data about structures of all these three bodies. Mars Express mission ended in 2014 (Wilson 2005).

- ROSETTA—One of the first deep space missions in world and the first one in Europe. The Rosetta was launched in 2004 and its main goal was flyby near the Churyumov-Gerasimenko comet. During the mission, the probe got as close as possible to two other asteroids called Steins and Lutetia (Ulivi and Harland 2015). After ten years from the launch, in 2014, Rosetta reached the main aim of the mission, and the lander Philae successfully settled on the comet. When the Rosetta was close enough, the lander was disconnected with the probe and dropped on the comet surface. Rosetta performed high resolutions surface maps and analyzed the asteroid and the comet structure. Lander Philae measured the comet structure and composition by drilling 20 cm deep hole in the surface. Lander obtained the probes and analyzed them in the onboard laboratory, sending data to the Earth. Over subsequent years, Rosetta flew with comet as her satellite and researched the processes occurring on its surface. The mission ended on 30th September 2016 and Rosetta fell on the Churyumov-Gerasimenko comet (ESA Webpage 2).
- VENUS EXPRESS—The first European mission to Venus, which was launched in 2005 by the Soyuz-Fregat launcher. The probe used instruments and technology from Rosetta, as well as from Mars Express, and was built only in 33 months. The mission's operational phase began on 3 June, 2016, and sent to Earth the most valuable data about Venus from all other Venus orbiters (except of Magellan mission). Venus Express provided information concerning atmospheric circulation and composition (Ulivi and Harland 2015). The mission helped to understand the evolution of the planet, the complex dynamics, and chemical interactions on the surface. Venus Express mission ended in 2014—after the propellant was exhausted, the orbiter fell to the surface of the planet (ESA Webpage 3).

The European Space Agency highlights a few basic directions of development of the space sector for the years 2016–2025 in the ESA Long Term Plan. Deep space exploration has qualified as a part of human spaceflight and robotic exploration program (ESA/C-91 2015). In the past, exploration of near Earth bodies and outer space beyond the Solar System was viewed exclusively as scientific mission, but there are now new perspectives and capabilities. For over fifty years, exploration missions have provided innovative solutions, products, and services for users on Earth. The challenges have been posed by deep space missions, offering the society enormous amounts of scientific and technological knowledge, which is significant for better understanding the planetary system. Global and European space exploration tendencies are as follows: establishing sustained access to destinations such as the Moon, Mars, and of course asteroids, and developing robots and human missions beyond Low Earth Orbit (LEO) (ISECG 2013). In addition, there are new partners, not only scientific institutions but also commercial companies interested in exploration due to the potential economic benefits.

Leading and significant role of space exploration missions have been underlined in new program, *Exploration Preparation, Research, and Technology Program* (ExPeRT) within *European Exploration Envelope Programme* (E3P). This program consists of all activities related to robotic and human exploration missions beyond LEO (Mars and Moon). The activities connected with space mining have been mentioned in commercial partnerships as part of E3P (ESA/PB-HME-17 2016). ESA has been recognizing that activities as strictly linked with commercial cooperation. One of the objectives of the ESA E3P project is bringing a sample from deep space exploration mission to Earth by 2030 (ESA/PB-HME-17 2016).

During the next decade, the European Space Agency is planning to continue ExoMars II, develop the Mars Sample Return (MSR) mission, and tighten cooperation with NASA. The first part of the MSR project, namely sending the Trace Gas Orbiter on the Red Planet's orbit, was completed within ExoMars Part 1. The next step is sending the orbiter (delivered by ESA) and lander (delivered by NASA) to Mars but still without samples. The full operational mission is being planned to be sent in 2026. European exploration strategy also assumes the development of Lunar exploration mission in cooperation with Roskosmos. ESA is also planning to realize activities connected with Moon Village building and utilization in situ resources. The European Space Agency for this purpose had been researching various resource extraction processes from materials available on the surface (ESA/IPC-69 2017). ESA in its short-term plan also mentions the International Space Station (ISS) and is going to continue the ISS Exploitation program and furthermore maximizes the usage of ISS and the output from the scientific experiment on the orbit. The European Space Agency also works on increasing strategic partnerships with the private sector in the context of ISS future utilization (ESA/C-91 2015). In 2019–2025, ESA plans on developing cooperation with China to maximize utilization of the ISS and Chinese Space Station. According to the *Space cooperation plan 2015–2020 between China National Space Administration and ESA*, which was updated in 2017 (ESA/IRC-10 2017), both agencies expressed their common interest in the human missions, infrastructure building, and scientific analysis of the lunar regolith (ESA/IRC-10 2017).

The European Space Agency has extensive experience in deep space exploration missions. The experience has been acquired by numerous mission beyond LEO. ESA is currently preparing programs for new and more sophisticated exploration missions such as Mars Sample Return, Phootprint mission or Lunar Sample Return, and in situ regolith utilization missions (ESA/IPC-69 2017).

4.2 Space Mining Trends in the United States

NASA is in the phase of conceptual projects and system testing designed to search, extract, and process minerals from asteroids, moons, and Mars. NASA focuses on the surface diagnosis regarding the presence of minerals. According to the *NASA Fiscal Year 2017 Budget Estimates*, the planned budget for 2017 includes, among other activities, placing ISRU devices for oxygen production on Mars and finalizing OSIRIS-REX and InSight missions (NASA Fiscal Year 2017). NASA's 2017 science budget includes the technologies needed to develop ISRUs. In situ resource utilization and asteroid redirect missions are outlined in the proposed program budgets: Space Technology Research and Development ($579.4 million) and Exploration Research and Development ($477 million USD) (NASA Fiscal Year 2017). In addition, both documents emphasize the possibility of reducing the cost of remote exploration missions by developing missions dedicated to asteroids.

NASA Technology Roadmaps, as one of the main initiatives, defines the asteroid and Mars missions and emphasizes the development of the technologies needed to achieve this goal. The subject of missions to asteroids or Mars occurs repeatedly in the context of manned missions and unmanned missions (NASA Technology Roadmaps 2015). Manned asteroid flight is scheduled for 2025 and 2030 for Mars. In the *NASA Technology Roadmaps 2015*, Human Exploration Destination Systems has repeatedly emphasized the need for the development of ISRU technology and has set goals for the coming years (NASA Technology Roadmaps 2015).

The International Space Exploration Coordination Group underlined in their report, the *Global Exploration Roadmap* (ISECG Roadmap 2013), that the common goals of the ISECG, which unite the national space agencies participating in the group, are identified in the following way:

- development of technology and infrastructure required for living and working beyond low Earth orbit;
- enhancing Earth's security by jointly developing a planetary defense system and mechanisms for managing space debris in orbit;
- increasing the human presence in the extraterrestrial space and increasing the level of human self-sufficiency in space missions;
- searching for extraterrestrial life.

Based on the study of the possibility of launching missions to Mars, the agencies continued to define the options that would be needed to perform manned missions beyond low Earth orbit. ISECG mission scenario (ISECG 2014) considered that in the longer term, the strategy could become more profitable by providing commercial services and applying public–private partnerships. By considering the planned and conceptual capabilities of the ISEGC Mission Scenario, the ISEGC mission identifies a set of missions in the orbit of the Moon and on its surface that will increase readiness for the Mars mission after 2030. The ISEGC mission scenario promotes increased human-robot collaboration through the integration of manned and unmanned missions (ISECG 2014).

The most promising ISRU mission of NASA is the RESOURCE PROSPECTOR mission. The mission is intended to be the first of its kind in terms of another celestial body extraction. The plan of the mission assumes the use of instruments capable of absorbing materials from the polar regions of the Moon. The purpose of the mission is to test solutions for extracting natural resources (mainly water, hydrogen, and other volatile substances) and to confirm their presence on the Moon. The mission assumes sending the rover, which will drill at a depth of 1 m below the surface, and then using a mounted experimental system (Regolith & Environment Science and Oxygen & Lunar Volatile Extraction), the sample is heated to determine the type and number of compounds present in the sample (such as hydrogen, nitrogen, helium, methane, and water).

It is necessary to emphasize that the American missions (LUNAR RECONNAISSANCE ORBITER and Lunar Crater Observation and Sensing Satellite (LCROSS), as well as NASA OSIRIS-REX) have increased the knowledge about the Moon and asteroids, which is technologically crucial for the future development. Lunar Reconnaissance Orbiter was launched into the lunar orbit in 2009 with the objective of mapping the Moon's surface, measuring radiation, searching for areas where the presence of ice is possible, and delivering images of the shadowed part of the Moon and terrain data regarding possible areas of landing. LCROSS, launched in 2009, has the objective of searching for ice water in permanently shadowed craters on the south pole of the Moon, determining the amount of water in the lunar regolith, and testing systems for subsequent space missions. As planned, a part of the LCROSS probe contacts the surface of the Moon. The second part of the probe performed measurements of the dust clouds generated by the impact and transmitted the obtained data to Earth. The objective of the third mentioned mission NASA OSIRIS-REX is to reach the Bennu (formerly 1999 RQ36) asteroid and to collect a 60-g soil sample, which later shall be transported to Earth and examined. The scientific objective of the mission is obtaining and analyzing the soil sample, identifying the mineral resources of the asteroid, measuring the Yarkovsky effect, and analyzing the regolith of the area where the soil has been collected. The mission was launched on September 8th, 2016 (Berry et al. 2013).

Currently, Asteroid Provided in situ Supplies is the only NASA concept mission that assumes the use of so-called optical mining technology. The technology is designed to collect large amounts of asteroid water (100 ton from one expedition). The mission assumes the use of sun-focused lenses and making use of such concentrated energy to make openings and extracting volatile material; evaporated water would be stored in the form of ice in passively cooled containers. Mining would take several months, and a portion of the water would be used as fuel for the Solar Thermal Propulsion drive system, which would help that the entire system could be moved to the orbit of the Moon.

The potential future mission proposed recently by NASA is still in the early stages of planning and development. ASTEROID REDIRECT MISSION with a planned launch date of December 2021 aims to rendez-vous with an asteroid and make use of robotic arms equipped in grippers or anchors to grab a 6-m boulder from the surface of the asteroid (NASA Technology Roadmaps 2015). The mission is designed to test the technology needed for manned flight. Moreover, the spacecraft would perform detailed observations and analyzes of the asteroid and would demonstrate at least one planetary defense technique.

In addition, private companies are strongly engaged in the process of making ISRU possible and facilitating the exploitation of extraterrestrial resources. If compared to Europe, progress of commercialization of space is very present in the United States. Policymakers and scientists have been thinking about the details of the commercialization of space for decades (Levine 1985). Commercial utilization of space is not only an essential component of services and market, but now the private sector's involvement in space is fast approaching a new paradigm thanks to a growing number of entrepreneurs, mostly in the United States (Velocci 2012). Involving the competition of private companies, the space industry will be built in new ways on the ground and in space. This new solution can create entirely new businesses (Anagnost 2017). The commercialization of space is the interest of venture capitalists who are looking for new business opportunities. Even NASA is interested in this process. Private sector companies will commodify technology currently used by the government and develop new solutions with either no or minimal government assistance. US industry entities performing operations in the field of space mining have been already investing in ISRU initiative. Asteroid robotic exploration is the long-term goal of Planetary Resources (formerly Arkyd Astronautic, est. 2009) entity. Currently, Planetary Resources is at the stage of creating and testing Attitude Determination & Control Systems, which is, as well as others, necessary for asteroid flights. Systems are being tested as satellite Arkyd series elements—the first one (Arkyd 3R) was launched in 2015 (Lewicki et al. 2013). The launches of two other Arkyd-6 spacecraft are scheduled for 2017. Other company Deep Space Industries (established in 2013) is planning at least two main missions: PROSPECTOR-X and PROSPECTOR-1. The launch of PROSPECTOR-X is planned during the 4th quarter of 2017 in Low Earth Orbit. The objective of the PROSPECTOR-1 mission is to leave the Earth's orbit and to come near to an asteroid. The mission shall test navigation systems and research instruments (i.e., neutron spectrometers capable of detecting water). By default, an attempt of landing on the surface and evaluating the features of the surface will be made. HARVESTOR-X and HARVESTOR-1 are planned missions with the same objective of exploring asteroids. Bigelow Aerospace Inc. (established in 1999) mainly focuses on habitat technologies development (Bigelow Expandable Module) on Earth's orbit, which may be used for conducting scientific research necessary for missions to the Moon and to Mars. Bigelow Aerospace bought the rights for lunar habitat technologies from NASA. Lastly, Moon Express (established in 2010) was founded in response to the Google Lunar X Prize contest, which had the objective of designing, building, and delivering to the Moon a rover capable of transmitting high-quality video of its Moon exploration progress to Earth. The Moon Express MTV-X1 rover was tested in December 2014, and Mars Express became the first enterprise to demonstrate a commercial lunar rover.

4.3 Chinese and Japanese Missions Related to Space Exploration

The Chinese space development plans in years 2016–2021 mainly concentrate on strengthening scientific exploration, technological innovation, institutional reforms; stimulating innovation and creativity; working to promote rapid development of the space industry; utilizing space resources in a prudent manner; and taking effective measures to protect the space environment to ensure a peaceful and clean outer space (China's Space Activities 2016). As China has the objective of developing its space activities mainly in space transport system and infrastructure, manned spaceflight, and deep-space exploration; experiments on new technologies (China's Space Activities 2016). China expand on the initiatives related to exploration of the Moon, which were commenced by the mission Chang'e 1 in 2007.

The summary of Chang'e program is listed as follows:

- Chang'e 1 (launch 2007) Chang'e 2 (2010)—the main task was to place the observation satellite in the orbit of the Moon;
- Chang'e 3 (launch 2013)—landing on the surface of the Moon;
- Chang'e 5 (launch 2017)—landing on the surface of the Moon, collecting lunar samples, and returning the samples to Earth;
- Chang'e 4 (launch 2018)—the first landing on the far side of the Moon, the exploration of the Moon by rover and establishing the launch a communication relay station to relay the signals between the lander/rover and the earth station at Earth-Moon L2 point in June 2018;
- Chang'e 6 (launch unknown)—unknown but performed within extended international cooperation (China's Space Activities 2016).

Additionally, China is planning to build the lunar base and to perform exploration of the polar regions of the Moon in the near future. On 11 January 2016, the Chinese government officially approved a robotic probe mission to Mars following 8 years of pre-research. The mission will launch a Mars probe, consisting of an orbiter, a lander, and a rover in the third quarter of 2020. The probe will conduct a soft-landing on Mars in 2021 and return to Earth with samples in 2030. According to China National Space Administration (China's Space Activities 2016), the probe will enter the Mars orbit and send the lander to soft-land on Mars surface and deploy the Mars rover to explore the surrounding area of the landing spot. The orbiter will continue orbiting Mars to survey its surface and provide data relay service for the lander and rover (China Space Report 2017). In addition, the Chinese space program contains a plan for exploring Jupiter, Venus, and asteroids (China's Space Activities 2016).

According to JAXA's Comprehensive Scenario for Space Exploration, Japan has developed an exploration plan of Mars and the Moon (JAXA 2015). The program towards exploration of the Moon highly depends on the success and outcome of the mission KAGUYA, which started in 2007, and the same year the observation satellite was placed in the orbit of the Moon. In June 2009, the mission KAGUYA has come to the end. KAGUYA consisted of the main orbiter and two small satellites: OKINA (relay satellite) and OUNA (VRAD satellite). The main orbiter was injected into a peripolar orbit of the Moon at an altitude of 100 km. The relay satellite was placed in an elliptic orbit at an apolune altitude of 2400 km to relay communications between the main orbiter and the ground station for measuring the gravity field of the backside of the Moon. The VRAD satellite measured the gravity field around the Moon by sending radio waves (JAXA 2015).

In the context of space mining, an exceptional achievement was gained due to HAYABUSA mission. For the first time, extraterrestrial matter from the planetoid has been delivered to Earth in 2010. In 2003, the mission MUSES-C (also known as HAYABUSA) was started. The space probe HAYABUSA reached S-type asteroid Itokawa using ion engines. It landed on the asteroid and gathered samples from two areas on the surface of the asteroid. The samples collecting system used explosives to launch the projectile that caused lifting of the dust from the asteroid surface. The dust was located in the specially prepared vessels, and in 2010, 1 g of the matter from the asteroid was delivered to Earth. In December 2014, HAYABUSA2 (the successor of HAYABUSA) was launched, and the spacecraft instruments and systems were, not considering small technical differences, basically similar to these used during the HAYABUSA mission. The spacecraft is equipped with a sampler mechanism, re-entry capsule, LIDAR, impactor, and Rover (Minerva-II) (Tsuda et al. 2013). HAYABUSA2 is intended to reach a C-type asteroid Ryugu to study the origin and evolution of the Solar System and materials for life in 2018. It is planned that the spacecraft will stay on the asteroid for approximately 18 months and return to Earth, carrying the samples in 2020 (Basic Plan on Space Policy Government of Japan 2013).

Both the mentioned missions are the part of program that explores S-, C-, and D-type asteroids. However, the target of the mission of asteroid explorer HAYABUSA2 is a C-type asteroid, which is a more primordial body than Itokawa (an S-type asteroid that is considered to contain more organic or hydrated minerals, although both, S- and C-types asteroids, have lithologic characteristics). Minerals and seawater which form the Earth as well as components necessary for development of life are believed to be strongly connected in the primitive solar nebula in the early Solar System.

According to Basic Plan and Space Policy 2016, the Japanese government plans to intensify the cooperation between Japanese private industrial companies (such as ispace, inc.) to conduct preparations for the next missions related to exploration of the Moon and, later, Mars. According to the press release of Japan Aerospace Exploration Agency (JAXA) (Akihiro 2016), Japan also plans to put a man on the Moon around 2030. Nevertheless, a spokesman of JAXA suggested that because of the high cost, JAXA is not planning to send an exclusively Japanese spacecraft to the Moon. Instead, the Japanese space agency prefers contributing to a multinational manned lunar mission. In Japan Fiscal Year 2019 (Tanaka 2017), the mission Smart Lander for Investigating Moon will be conducted. Japan is going to develop the probes that can perform the sample return mission from the moon of Mars in early 2020 and conduct the lunar Antarctic exploration in the same period. The next mission regarding the Moon exploration program envisages the stay of 4 men crew for 500 days in the low-gravity environment of the Moon. During the mission, which will occur in 2030, the crew will perform the trial of fuel production. The next step of the Mars exploration program will be sending an unmanned probe that will gather the information about the environment and make the basic investigation regarding the future manned mission, which will occur in 2045.

4.4 Russian Space Exploration Program

The State Space Corporation, Roscosmos was re-established in 2015 and has taken over the Russia Federal Space Agency responsibilities. In Russia, three main planetary exploration programs have been developed: Lunar, Martian, and Venusian. During the Lunar Exploration Program, called Luna, which aim was researching the Moon surface and bringing samples of the Moon regolith to Earth, 24 vehicles (landers, rovers, and probes) were sent. From all the missions, 20 ended with success (Williams 2005).

In 1961, within the Venera program, Russia launched the first Venusian mission. The program was prepared to study the surface and the atmosphere of the planet, including in situ experiments, landers, and orbiters. Missions delivered comprehensive studies about clouds, greenhouse effects, and volcanic processes on Venus. The next step in Venus exploration was the Vega series. Missions include Venus and Comet Halley flybys, which Russian participated in the ESA Venus Express mission. For example, Roscosmos developed and participated in 32 Venusian missions, from which 16 succeeded (Zelenyi 2016). However, the Mars exploration program was not as successful compared to the Russian contributions to lunar (Luna and Lunokhod), Venus, and comet explorations (Venera, Vega). Only 1 of the 20 missions that were developed by Russia or in which Russian scientists participated were successful (Zelenyi 2016). Phobos and Mars mission series failed generally because of spacecraft signal failures, onboard computer errors, and trajectory complications.

The satellite of Mars, Phobos, was not reached by Russian missions (failed missions: Phobos in 1988, Phobos Sample Return in 2011). However, Phobos is still perceived as an interesting target and probably a key to the history of the early Solar System. The ultimate goal is to bring back a sample of Phobos regolith and study it in the laboratory on Earth. It is worth mentioning, that the Polish subsurface instrument CHOMIK was part of the onboard systems of Phobos Sample Return mission.

In the future, Russian space policy orbital constellation and launcher development as well as continuation of Lunar and Mars exploration and ISS utilization have been scheduled. Near-term missions are planned in partnership with ESA (ExoMars) and as a continuation of Luna Series. Roscosmos plans to develop technologies for Luna-Grunt sample return (launch planned in 2021), Boomerang Phobos Sample Return (2022), and Mars sample return (2024). That ambitious objective can be achieved by enhancing the impact of its resources through partnerships with other space agencies. Federal Space Program is considering as an important factor cooperation with private sector (Pischel 2017).

5 Potential and Expertise of Polish Entities—Achievements and Prospects for Further Development

5.1 Polish Expertise and Space Heritage Related to Space Mining

Polish entities, especially Space Research Centre of the Polish Academy of Sciences (CBK PAN), accompanied by other stakeholders at the Polish scientific and industrial market, have been developing since years of strategic projects related to subsurface exploration in microgravity conditions and sampling of regolith. The most prestigious and relevant technology development instruments and devices are listed as follows:

- MUPUS—the MUPUS Instrument for Rosetta mission to Churyumov-Gerasimenko comet was developed by CBK PAN on behalf of the German Aerospace Center (DLR) and ESA. Multi-purpose sensor package onboard the Rosetta lander Philae was designed to measure the energy balance and the physical parameters in the near-surface layers—up to about 30 cm depth—of the nucleus of Rosetta's target comet Churyumov-Gerasimenko and monitor changes in these parameters over time as the comet approaches the Sun. Among the studied parameters are the density, the porosity, cohesion, the thermal diffusivity and conductivity, as well as temperature. Polish contribution to the experiment was substantial, including manufacturing and technical qualification of the instrument. MUPUS was meant to work in extreme conditions, such as in a vacuum, with high temperature gradients, with resistance to huge overload (even required 1000 g), and have the expectation of good reliability after several years of space travel. The technical objectives were achieved with very low resources exploitation of resources: 1500 g of weight and about 2 W power supply. MUPUS and its deployment of the robotic system allowed the transfer of thermal sensors into regolith, measurement of thermal properties of the soil, and measurement of mechanic properties of the regolith (Spohn et al. 2007; Kömle et al. 2002; Grygorczuk et al. 2007, 2008, 2009a).
- KRET—mole device was constructed by CBK PAN and was one of the solutions for low power consuming device that could be treated as a sub-surface end-effector of a more complicated robotic system. The detailed investigation of lunar analogue showed how the progress of the mole depends on the compaction ratio of the regolith analog. Prototypes of mole type penetrators dedicated for Moon investigation (KRET) were demonstrated with a proposition of its deployment system—the Ultra-Light Planetary Manipulator, also designed by

CBK PAN. In addition, the KRET penetrator is equipped with several detectors, especially for thermal measurements (Grygorczuk et al. 2009b, c; Seweryn et al. 2013, 2014b).

- CHOMIK—the unique geological device, CHOMIK penetrator, dedicated for the Russian Phobos Sample Return space mission was constructed in 2010 by CBK PAN. One of the key goals of the mission was to acquire a regolith sample from Phobos, Mars' moon, and to deliver it to Earth. The sample was to be collected from the surface of the moon by the Polish penetrator and to be deposited in a container. However, the mission did not succeed (Gurgurewicz et al. 2010; Rickman et al. 2014; Seweryn et al. 2014b).
- SPACE DRILLING MECHANISM—the project "Development of the automatic core drill working in extreme conditions, especially space environment" was financed by the Polish National Centre for Research and Development (NCBR). It was executed by the consortium with CBK PAN as the leader and two faculties of AGH University of Science and Technology (Faculty of Drilling, Oil and Gas and Faculty of Mechanical Engineering and Robotics, Department of Robotics and Mechatronics) as partners. CBK PAN has been working on the development of autonomous core drill for extreme environments since 2012 when the project was initiated. In 2014, prototypes of the key modules of the system were manufactured and assembled. In 2015, the prototypes of the key modules were separately tested to verify the compliance with the requirements and validate required functionality, e.g., in vacuum. Finally, the whole system was assembled and tested in the newly constructed testing facility for subsurface operations in CBK PAN. The aim of the project was to design and construct a device that could be used for drilling on planets and other celestial bodies or in harsh terrestrial conditions. It is easy to move and transport, as the device has the self-automated driving system (Seweryn et al. 2014a).
- MOON REGOLITH ANALOGUE—thanks to the cooperation between AGH University of Science and Technology and CBK PAN, the analogue of the Moon ground was produced in 2014. This substance was created based on the very exact data about geo-mechanical features of an analogue produced in the United States (Bednarz et al. 2013; Heiken et al. 1991; Luding 2004; Seweryn et al. 2008).
- SAMPLER—Sample Acquisition Means for the Phootprint Lander: Experiments and First Realization, the project is performed by CBK PAN on behalf of ESA; its aim is to test the influence of regolith sampling devices on the stability of a lander in the environment of low gravity bodies.
- PACKMOON—the main contractor of this project is CBK PAN. It was launched in the first quarter of 2015, and the end is scheduled for the last quarter of 2017. The aim of the undertaking is to develop a regolith sampling device dedicated to low gravity bodies. This device will operate based on two assumptions: drilling of hammering elements into regolith and minimal interaction with a lander by doubling the system (hammering mechanism, casing and backup mass) (Seweryn 2016).

- HP3—Hammering Mechanism for DLR's HP3 penetrator *NASA InSight mission* (2013)—for NASA mission InSight, Astronika, Warsaw University of Technology and CBK PAN (subcontractors) in cooperation with DLR (German Aerospace Center) developed a hammering mechanism for the HP3 penetrator, which was constructed by DLR. The main task of HP3 is to use an electromechanical impact mechanism capable of driving an instrument container into the Martian surface to a depth of up to five meters. This hammering mechanism will be equipped with active and passive detectors enabling the exact assessment of the regolith's features (Astronika Webpage).
- LUNAR DRILL DEVELOPMENT—the project was initiated in 2015 and is performed by the Polish company Astronika (subcontractor) in cooperation with the Italian company SELEX (prime) in frame of ESA program—GSTP. This instrument is dedicated to the Russian-ESA Moon Luna-Resurs (Luna-27) mission. The aim of the mission is to become familiar with the Moon environment, as well as to examine the possibility of the usage of the Moon's regolith. As a result, a regolith sampling device will be developed. It will enable drilling up to two meters below the surface and taking samples.
- ROBOTICALLY-ENHANCED SURFACE TOUCHDOWN—the project was launched in 2015 and is presently performed by the international consortium in which Polish entities are participating: GMV Innovative Solutions and CBK PAN. Its aim is to design and develop an active landing gear system in the ESA mission Phootprint. The mission is a part of ESA's *Mars Robotic Exploration Programme*. Sample return from Phobos is one of the main scientific goals of this mission (Barraclough et al. 2014).

5.2 *Main Polish Stakeholders Involved in Space Exploration*

Polish competencies are based on several entities involved in the cooperation with both the European Space Agency and the European Commission. The majority of them are small and medium sized enterprises accompanied by prestigious research institutes such as the Space Research Centre of Polish Academy of Sciences (CBK PAN), Industrial Research Institute for Automation and Measurement (PIAP), and technical universities in Warsaw, Lodz, Cracow, Poznan and Wroclaw.

Polish companies and research institutes are involved in prestigious, international space missions, such as Rosetta. For purpose of this mission, the hammering instrument called MUPUS was constructed by CBK PAN (Spohn et al. 2007; Kömle et al. 2002; Grygorczuk et al. 2007). The second program was the Russian Phobos-Grunt Sample return mission. Poland participated in it by creating a subsurface penetrator called CHOMIK (Gurgurewicz et al. 2010; Rickman et al. 2014). In addition, Polish stakeholders are involved in important ESA programs and missions implemented currently such as ExoMars, E.Deorbit within the Clean

Space Initiative, or Lunar Lander. They also expressed their interest in participating in the new ESA and NASA mission, called Asteroid Impact Mission, which has been presented by ESA 2016 (ultimately, the mission was postponed). Thus, they cooperate very closely with European large systems integrators (i.e., Airbus Defence & Space, Thales Alenia Space, and OHB). Poland also subscribed to optional programs concerning space robotics (Mars Robotic Exploration Preparation Programme and currently E3P).

There are two main competency areas of Polish entities involved in space robotics:

(1) Subsurface exploration: sampling devices, mechanisms working in dirty vacuum environments, subsurface navigation, control systems/control electronics, sensors, civil engineering, and mining in extreme environment.
In addition, the selected, active representatives of entities developing and implementing technologies at the Polish market can be mentioned: research institutions—CBK PAN and Industrial Research Institute for Automation and Measurement PIAP, and companies—Astronika, Creotech Instruments, Robotics Inventions, SENER Polska.
(2) Components for on-orbit robotic systems: guidance navigation and control systems, hardware components (e.g. manipulator joints or links, nets), motion sensors, and deployable structures.
As the selected, active representatives of entities developing and implementing that technologies at the Polish market can be mentioned: research institutions—CBK PAN, PIAP, Cilium Engineering, CIM-mes Projekt, GMV Polska, SKA Polska, OptiNav, SYBILLA Technologies, ABM Space.

The Polish entities are developing several projects in the area of space mining and planetary exploration on behalf of the European Space Agency (Table 1).

Polish entities from the space sector, especially those dealing with robotic exploration, were attracted by the Luxembourg Government initiative. The Luxembourg Government together with Deep Space Industries and the Société Nationale de Crédit et d'Investissement the national banking institution in Luxembourg, have signed an agreement formalizing their partnership to explore, use, and commercialize space resources as part of Luxembourg's spaceresources.lu initiative (Rausch 2016). The Luxembourg Government is willing to invest 200 million Euro for the purpose of research and development of technologies related to space mining (Bartunek 2016). Polish companies treat this initiative not only as a chance for their long-term presence in space, but also for performing specific business projects concerning participation in future exploitation of asteroids' resources. Hence, several of them created the Polish-Luxembourg dedicated consortium that will examine the perspectives of a future cooperation with Luxemburg, its strategic investors, and partners from the United States of America. Members of the consortium are among others: ABM Space, Astronika, GMV Innovating Solutions, PIAP, Hertz Systems, Creotech Instruments, and Sybilla Technologies (PAP Nauka w Polsce 2017).

Table 1 List of projects conducted by the Polish entities on behalf of the European Space Agency in the area of space mining and planetary exploration; state of the art: 29.12.2016

Name of the Polish entity	Title of the project	Name of ESA program	Start
ABM SPACE	Docking Impender	Basic Technology-Startiger	2014
Astronika	System Analysis of Deployable Components for Micro Landers in Low Gravity Environment	General Studies	2016
Astronika	Phase B+ of Prospect Development for Lunar Exploration	GSTP Period 6	2016
CBK PAN	Sample Acquisition Means for the Phootprint Lander: Experiments and First Realisation (Sampler)	AURORA MREP-2	2013
CBK PAN	Robotically-Enhanced Surface Touchdown (Rest)	AURORA MREP-2	2013
CBK PAN	1 kg Sample Acquisition System For Fast Surface Sampling	Polish Industry Incentive Scheme	2013
CBK PAN	LOOP—Landing Once on Phobos	Polish Industry Incentive Scheme	2016
GMV Innovating solutions	Guidance, Navigation and Control (GNC) for Phootprint Descent and Landing	AURORA MREP-2	2013
Sener	Umbilical Release Mechanism	AURORA MREP-ExoMars	2014

6 Industrial Policy and Legal Aspects

6.1 Intellectual Protection in the Field of Space Mining

Strategies of intellectual protection management used to obtain a temporary monopoly of a solution are not always applicable or have economical justifications for space solutions. There are patents applications and patents obtained for the invention of the broadly defined field of space research and exploration; however they do not constitute a common form of protection of intellectual property. This is due to a number of fundamental reasons, such as the uniqueness of the solution, the lack of economic legitimacy of space technology inventions, patent protection, significant know-how, and the key importance of space heritage. It is well known that in the space industry, space approval is the ultimate technological verification; therefore, it is important to demonstrate experience in product manufacturing and instrument use. For the space sector members, acquiring space heritage is a measure of recognition, success and a reference of development level.

In addition, acknowledging the importance and future prospective use of the designed space devices, the patent protection procedure was conducted. There are various routes to patent protection, and the selection of the optimal one depends on the specific invention and the markets of the company. Considering the international character of space market, main space Polish players are not only seeking protection in several countries, (they may apply directly for a national patent to each of the national offices), but they are also turning to the European Patent Office (EPO), with applications under the European Patent Convention and the Patent Cooperation Treaty. Currently, the number of patent filings regarding space oriented inventions under Polish Patent Office is significantly higher than under EPO. The most important European patent applications and patents obtained in the field of space mining are as follows:

- EP3087244—System for forming drill tube from flat strip wound on a drum and drilling method using that tube, CBK PAN and AGH.
- EP3066716—Clamping mechanism, locking arrangement, and method of operating a reconfigurable manipulator, CBK PAN.
- EP3063776—Electromagnetic drive and method of production thereof, CBK PAN.
- EP3093427—Drive for tubular member, curling strip, and tubular boom, CBK PAN.
- EP3057894—Mechanism, method, and escapement wheel for controlling rotational motion, CBK PAN.

Table 2 presents the list of the Polish patent applications related to space mining made under the Polish Patent Office.

6.2 International Law Aspects

Space tourism, space mining, exploration of the edge of the Solar System, space commercialisation trends, VC interest of space activities—that all facts influence the urgent necessity for the establishing new law more adequate to the technology development and human ambitions. The law of outer space has addressed the new political and economic needs that meet the human will to go into space. Since a few decades, successful exploration mission has been creating the possibility to explore new planets. Building of settlement and habitats on the Moon will come true. International law is lacking the regulation concerning new challenges; it is not responding to the needs of societies and the dynamics of progress in the space exploration (Tronchetti 2009).

The United Nations Committee on the Peaceful Uses of Outer Space (UN COPUOS) is responsible for the international space law construction and has been negotiating the most important treaties, listed below (according to international law the treaties signed by a Party State means the government of the country is aware of the fact, that the treaty exists, but it does not mean that the country is accepting the treaty and is not forced to include the obligations consequent to the treaty in its national regulations. If the document has been ratified by the country, it means that

Table 2 Patent applications related to space mining made under the Polish Patent Office

Applicant/Polish entity	Application number	Date of application	Publication date	International patent classification code	Invention title
Astronika	407198	16.02.2014	17.08.2015	E21C51/00 G01N1/08 B64G1/66	Method for Sticking into the Ground Subsurface Layers the Device for the Sticking into the Ground by this Method
Astronika	408263	19.05.2014	23.11.2015	E21B4/04 H02K1/00 H02K37/00	Method for Producing a Torque, Preferably for Drilling Rigs and the Driving Device for Pulsed Production of the Torque
CBK PAN Akademia Górniczo-Hutnicza Im. S. Staszica in Kraków	406633	23.12.2013	06.07.2015	E21B7/00 E21B17/00 E21B19/22 F16L9/16	System for Forming Drill Tube From Flat Strip Wound on a Drum and Drilling Method Using That Tube
CBK PAN	405646	15.10.2013	27.04.2015	F16H27/02	Mechanism, Method and Escapement Wheel for Controlling Rotational Motion
CBK PAN	405821	29.10.2013	11.05.2015	H02K33/02 G01N1/08	Electromagnetic Drive and Method of Production Thereof
CBK PAN	405939	06.11.2013	11.05.2015	F16L33/02 F16L33/20	Clamping Mechanism, Locking Arrangement and Method of Operating Reconfigurable Manipulator
CBK PAN	407697	28.03.2014	12.10.2015	E04C3/00 B65D85/67 B64G99/00	Method for Uncoiling of a Tube Wound from a Tape and the Mechanism of the Wound Tape
CBK PAN	406472	11.12.2013	22.06.2015	B65H75/34	Drive for Tubular Member, Curling Strip and Tubular Boom
CBK PAN	397651	30.12.2011	08.07.2013	G01N33/24	Lunar Soil Analogue
CBK PAN	401280	19.10.2012	28.04.2014	B65H54/20 B25J18/02	Drive for Rolled Tape and the Manipulator

(continued)

Table 2 (continued)

Applicant/Polish entity	Application number	Date of application	Publication date	International patent classification code	Invention title
CBK PAN Gut Henryk Gutronic	403660	25.04.2013	27.10.2014	B21C37/08 B21D5/10 F16L9/16	Device for Producing Tubes of the Elastic Strips, the Method for Producing Strips of Elastic Tubes and a Tube Formed from Elastic Strip
PIAP	411601	15.03.2015	28.04.2017	G05D3/14 G05B11/28 G05B11/42	Method for Controlling a Direct Current Motor
PIAP	411600	15.03.2015	28.04.2017	G05B11/42 G05D3/14 G05B11/28	Control System for a Direct Current Motor
PIAP	414344	12.10.2015	–	G01R33/07 H01L43/06	Magnetic Field Sensor System
PIAP	414129	24.09.2015	–	H01L27/22 H01L43/06 G01R33/07	Hall Generator with Offset Compensation

the treaty, and obligations consequent to it, was accepted by the Parliament of the country and the government of the country is going to submit to the treaty):

- *Outer Space Treaty* (*Treaty on Principles Governing the Activities of States in the Exploration and Use of Outer Space, including the Moon and Other Celestial Bodies*)—entered into force on October 10, 1967; it was signed by 25 countries and ratified by 105 countries. Outer Space Treaty concerns: collective benefit of the exploration and use of outer space for all mankind; freedom to explore outer space regardless of economic or scientific development of the country; ban of placing in the Earth orbit of the weapons of mass destruction; international liability of launching State; jurisdiction and control regarding the object of the State which registered it (Outer Space Treaty 1967).
- *Rescue Agreement* (*Agreement on the Rescue of Astronauts, the Return of Astronauts and the Return of Object Launched into Outer Space*)—came into force on December 3, 1969; signed by 24 countries and ratified by 95. Its main provisions concern: obligation to notify the launching authority and the Secretary-General of the UN about dangerous conditions of the personnel of a spacecraft of the foreign launching authority; obligatory rescue of the personnel of a spacecraft by the country, where the accident landing occurred; obligation to return the space object and its components to launching authority on expense of the launching authority (Rescue Agreement 1969).
- *Liability Convention* (*Convention on International Liability for Damage Caused by Space Objects*)—entered into force on September 1, 1972; it was signed by 20 countries and ratified by 94. That convention includes provisions regarding: absolute liability for compensation of damage and injuries caused by the space object of a country in the atmosphere of Earth; fault of the country must be proven in case of liability for damages and injuries caused elsewhere (Liability Convention 1972).
- *Registration Convention* (*Registration of Objects Launched into Outer Space*)—came into force on September 15, 1976; it was signed by 4 countries and ratified by 63. It concerns: obligation of carrying by a State a Register of the space objects owned by the State and informing of Secretary-General about entries made into the Register; information regarding the object which shall be furnished to the Secretary-General of the UN and obligation of marking the space objects with a designator (Registration Convention 1976).
- *Moon Agreement* (*Agreement Governing the Activities of States on the Moon and Other Celestial Bodies*)—came into force on July 11, 1984; it was signed by 4 and ratified by 17 countries. Moon Agreement emphasizes: exclusively peaceful purpose of the Moon; the exploration and use of the Moon as the province of all mankind; freedom of scientific investigation on the Moon; the Moon and its natural resources as the common heritage of mankind and not the subject to national appropriation (Moon Agreement 1984).

The *Outer Space Treaty* is the main treaty regulating the fundamental legal issues related to the exploration of outer space and celestial bodies. According to the Treaty, exploration of outer space shall be carried out for the benefit and in the interests of all countries, irrespective of their degree of economic or scientific development, and shall be the benefit of all mankind. It is also mentioned that *Outer Space*, including the Moon and other celestial bodies, shall be free for exploration and use by all States without discrimination of any kind, on a basis of equality and in accordance with international law, and there shall be free access to all areas of celestial bodies. In article 2 *The Outer Space Treaty* mentions that *Outer Space* is not subject to national appropriation by claim of sovereignty, by means of use or occupation, or by any other means.

Moon Agreement regulates issues related to mining of resources on celestial bodies. The Treaty "does not apply to extraterrestrial materials which reach the surface of Earth by natural means" (Moon Agreement 1984). Moreover, the Moon Treaty restricts the term of concerned celestial bodies to these which are located in the area of the Solar System (except Earth). From the legal point of view the difficult issue related to space mining comes from the Moon Treaty (art. 4 § 1): The exploration and use of the Moon shall be the province of all mankind and shall be carried out for the benefit and in the interests of all countries, irrespective of their degree of economic or scientific development. Due regard shall be paid to the interests of present and future generations as well as to the need to promote higher standards of living and conditions of economic and social progress and development in accordance with the Charter of the United Nations. Two interpretations of this provision exist. The first recognizes the virtual interpretation of the *Moon Agreement* and anticipates necessity of division of resources obtained on the celestial bodies, and impossibility of mining resources on the celestial bodies except acquiring them exclusively for scientific purposes (in this situation: results of researches have to be conveyed *mutatis mutandis* as in the case of resources mentioned above). The second, adopts the interpretation of the *United Nations Convention on the Law of the Sea (UNCLOS)* (art. 87 § 1) (United Nations Convention on the Law of Sea 1982) which entitles to freedom of fishing on the Open Sea and freedom of scientific research on the Open Sea. Literally, the *Moon Agreement* does not concern the issue of mining resources on asteroids or other natural sources.

There is a need to consider in law at least five issues relating to utilisation of space resources:

- rules of international cooperation in space;
- rules of international competition;
- space mining property rights and intellectual property issues;
- legal liability;
- consequences for the environment (contamination risk and protection needs).

According to the most presumable scenario of development of international law in the field of space mining, necessity of creating different solutions regarding this subject will force international society and the United Nations to work out the new provisions or complement the old ones related to mining of resources in space environment. As a result of commercialization of outer space, it might be expected that in the future new provisions related to definition of property rights in outer space will be constituted (Tronchetti 2009). Furthermore, because of the fact that legal definition of celestial body does not exist, it is impossible to adopt any regulations in regard of space mining. The necessity of creating national space law regarding space mining may only occur in the situation, when in one country would exist few entities able to perform space mining on celestial body. The extremely important branch of law regarding space mining is definition of property laws in executive arrangements signed with other participants of joint space missions. In case of individual missions the meaningful is the fact, that the *Moon Agreement* was ratified by 17 and signed by 4 countries and its provisions are not significant for a big group of countries.

Commercialization of space activities is the new domain of space exploration, and thereby pose questions for the regulation and the use of the most densely populated orbits (IISL 2016). On 25th November 2015, the president of the United States of America signed the *Space Resource Exploration and Utilization Act of 2015* which is the part of the *U.S. Commercial Space Launch Competitiveness Act.* The American national legislation related to space activity entitle an American company to possess, own, transport, use, and sell the space resources obtained through its commercial undertakings. The Act underlines the fact, that "the undertakings of the private companies shall be performed in manner consistent with the existing international obligations of the United States", what implies that the national law under which licenses for space mining exists. Presently, the U.S. government did not issue such a regulation. The *Space Resource Exploration and Utilization Act of 2015* does not allow the American company to occupy or possess the space object but to possess, transport, process and sell the resources obtained in outer space.

Furthermore, Luxemburg, as the first European country, decided to determine the legal framework for the space explorations, and especially related to the initiative of space mining. The Luxembourg Government envisages two options: either to adopt the so far existing space law from 1960, or to create a set of new acts and law regulations. The law to be adopted should assure private investors the guarantee that their activities are legal, safe, and the right to exploit and use of cosmic raw materials from asteroids will be protected (Luxembourg to launch framework to support the future use of space resources 2017).

The establishment of national law regarding space mining arose a dynamic dispute in UN COPUOS where the points of view on this issue are very divided. The representatives of the first interpretation of the outer space Treaty emphasize that internal regulations regarding space mining can only be respected by the entities registered in this country and are non-binding for citizens of other nationalities. Some of the representatives are also aware of the fact, that national space activity acts do not consider the threat to the environment of Earth which may bear resources obtained in outer Space. Therefore, in view of the absence of a clear prohibition of extracting resources in the *Outer Space Treaty* one can conclude that the use of space resources is permitted (IISL 2015). The new United States Act can be treated as a possible interpretation of the *Outer Space Treaty*. Whether the United States' interpretation of Art. II of the Outer Space Treaty will be followed by other states will be crucial to the future understanding of the non-appropriation principle. As *Position Paper On Space Resource Mining* states, "It can be a starting point for the development of international rules to be evaluated by means of an international dialogue in order to coordinate the free exploration and use of outer space, including resource extraction, for the benefit and in the interests of all countries" (IISL 2015).

Summarizing, the international community is divided between those who state that national space activity acts shall be regarded as the source to introduce the amendments to the international space law which is now based solely on the five UN space treaties; and those who are adherents of the statement that all of the UN space treaties shall be absolutely abided by the whole international community as the acts of peremptory norms (*ius cogens*). It is expected, that in the next few years there will be several bilateral or multilateral agreements that will settle on an international convention, similar to how deep-sea mining is handled today. Problems which need urgent attention are in particular: the complex problem of space debris, commercialization of space, space traffic management, the issue of attribution of liability and, the need for the inclusion of insurance policies in the existing legislation and space mining limits or permissions (IISL 2016). Shaping new legal rules for space exploration, especially in context of ISRU is desirable.

7 Conclusions

The sampling of materials from celestial bodies and transporting them to Earth requires a systematic development of extensive industrial and technological competences on the various stages of space mining process, such as mission planning and its components, mobility systems, robotic systems, land acquisition tools (drilling, prospecting, gathering, packaging), and finally, analytical and research systems. Nowadays, missions beyond the Earth orbit are not affecting only knowledge benefits, also the economic increment is one of the exploration programs goal. Technology development process related to space mining can create an unique platform of international cooperation for the Polish companies. Moreover, it can strengthen efforts aimed at preparation for future deep space exploration missions during which Polish devices and inventions will be used. Future advances in space mining technology can push towards the larger goal of exploration in extraterrestrial environments, and this aim facilitates Polish R&D participation in the global space supply chain. It can also create on the Polish market a new field of activities for sectors and companies that are not involved in space activities. Nevertheless, Polish firm competences and previous experiences in space exploration show that they have strong abilities and that it is highly possible that they will become important players in space mining activities within the next 10 years.

The contracts received from ESA by the Polish firms has contributed to the improvement of their abilities and technologies in the area related with space mining.

The Polish Space Agency's (POLSA) mission is to support innovative and visionary industrial ideas. POLSA has ambitions to support Polish entities in the field of widely defined space mining mainly due to the very broad context of technological development related to surface and subsurface deep space exploration. POLSA will actively support space mining undertakings through governmental funding. The National Space Program is being designed to implement the Polish Space Strategy, which aims at increasing the competitiveness of Polish space technologies and satellite applications oriented companies. Space mining can also be included among the major Polish specialisations. Active involvement of public entities and large state-owned companies in R&D projects is extremely desirable—POLSA will actively participate in these activities through creating favorable conditions for the development of such undertakings.

References

Akihiro I (2016) Establishing deep space exploration technology and new challenges. JAXA. http://global.jaxa.jp/projects/sat/HAYABUSA2/. Accessed 4 Aug 2017

Anagnost A (2017) Commercialization of space: 4 predictions for the 21st-century gold rush. https://redshift.autodesk.com/commercialization-of-space/. Accessed 3 Aug 2017

Astronika Webpage. http://www.astronika.pl/hp3-mole. Accessed 26 July 2017

Bar-Cohen Y, Zacny K (eds) (2009) Drilling in extreme environments: penetration and sampling on earth and other planets. Wiley-VCH Verlag GmbH & Co. KGaA, Weinheim

Barraclough S, Ratcliffe A, Buchwald R, Scheer H, Chapuy M, Garland M, Rebuffat D (2014) Phootprint: a European phobos sample return mission. In: Proceeding of the 11th international planetary probe workshop IPPW, California Institute of Technology, Pasadena, California, 16–20 June 2014

Barton A (2015) Update on the Google Lunar XPRIZE. Paper presented at the MOON 2020-2030, ESA ESTEC, Noordwijk, 14–16 Dec 2015

Bartunek RB (2016) Luxemburg sets aside 200 million euros to fund space mining ventures. Reuters. http://www.reuters.com/article/us-luxembourg-space-mining-idUSKCN0YP22H. Accessed 27 July 2017

Basic Plan on Space Policy (2013) Strategic headquarters for space policy. Government of Japan

Bednarz S, Rzyczniak M, Gonet A, Seweryn K (2013) Research of formed lunar Regholit analog agk-2010. Arch Min Sci 58(2):551–556

Berry K, Sutter B, May A, Williams K, Barbee BW, Beckman M, Williams B (2013) OSIRIS-Rex Touch-And-Go (TAG) mission design and analysis. In: Proceedings of the 36th annual AAS guidance and control conference, Breckenridge, Colorado, 1–6 Feb 2013

Blair BR, Gertsch LS (2010) Asteroid mining methods. Paper presented at the SSI space manufacturing 14 conference, NASA Ames Research Center, 29–31 Oct 2010

Całus Moszko J, Białecka B (2012) Potencjał i zasoby metali rzadkich w świecie oraz w Polsce. Prace Naukowe Głównego Instytutu Górnictwa – Górnictwo i Środowisko. Kwartalnik 4: 61–72

Carrier WD (2003) Particle size distribution of lunar soil. J Geotech Geoenviron Eng ASCE 129 (10):956–959

Carrier WD, Mitchell JK (1989) Geotechnical engineering on the Moon. Bechtel Inc. and University of California. https://ntrs.nasa.gov/archive/nasa/casi.ntrs.nasa.gov/19910073656.pdf. Accessed 9 Aug 2017

Carrier WD, Olhoeft GR, Mendell W (1991) Physical properties of the lunar surface. In: Heiken D, Vaniman D, French BM (eds) Lunar sourcebook. Cambridge University Press, Cambridge, pp 475–594

Chamberlain PG, Taylor LA, Podnieks ER, Miller RJ (1993) A review of possible mining applications in space. In: Lewis JS, Matthews MS, Guerrieri ML (eds) Resources of near-earth space. The Arizona Board of Regents, Phoenix

China's Space Activities in 2016 (2016) III Major tasks for the next five years. The State Council Information Office of the PRC

China Space Report (2017) Mars exploration mission. China Space Report. https://chinaspacereport.com/spacecraft/mars-mission/. Accessed 4 Aug 2017

Cohen MM (2013) Robotic asteroid prospector (RAP) staged from L1: start of the deep space economy. NASA Innovative and Advanced Concepts NIAC. https://www.nasa.gov/pdf/740784main_Cohen_Spring_Symposium_2013.pdf. Accessed 20 July 2017

Colwell JE, Batiste S, Horányi M, Robertson S, Sture S (2007) Lunar surface: dust dynamics and regolith mechanics. Rev Geophys 45(2)

Crites ST, Quintana S, Przepiórka A, Santiago C, Trabucchi T, Kring DA (2012) Lunar landing sites that will enhance our understanding of regolith modification processes. Paper presented at the 43rd lunar and planetary science conference, The Woodlands, Texas, 19–23 Mar 2017

Davis BH, Occelli ML (2010) Fischer-Tropsch synthesis, catalysts, and catalysis: advances and applications. CRS Press, New York

ERECON Report—European Rare Earths Competency Network (2014) Strengthening the European rare earths supply-chain: challenges and policy options. http://reinhardbuetikofer.eu/wp-content/uploads/2015/03/ERECON_Report_v05.pdf. Accessed 3 Aug 2017

ESA Webpage 1. http://www.esa.int/Our_Activities/Space_Science/SMART-1. Accessed 25 July 2017

ESA Webpage 2. http://rosetta.esa.int/. Accessed 25 July 2017

ESA Webpage 3. http://sci.esa.int/venus-express. Accessed 25 July 2017

ESA/C-91 (2015) European space agency/council, Draft ESA Council Long-Term Plan 2016–2025, Annex 1. ESA

ESA/IPC-69 (2017) European space agency/industrial policy committee, add.1. ESA

ESA/IRC-10 (2017) European space agency/international relations committee, China—Outcome ESA/CNSA Steering Committee. ESA

ESA/PB-HME-17 (2016) European space agency/human spaceflight, microgravity and exploration programme board (2016), Draft European Exploration Envelope Programme Proposal, Annex 5. ESA

Gambogi J (2017) Rare earths. In: Mineral commodity summaries. U.S. Geological Survey, Reston, Virginia, pp 134–135

Gaubert F (2015) BB 12: habitation systems. Paper presented at the MOON 2020-2030, ESA ESTEC, Noordwijk, 14–16 Dec 2015

Grygorczuk J, Banaszkiewicz M, Seweryn K, Spohn T (2007) MUPUS insertion device for the Rosetta mission. J Telecommun Inf Technol 1:50–53

Grygorczuk J, Seweryn K, Wawrzaszek R, Banaszkiewicz M (2008) Insertion of a mole penetrator —experimental results. In: Proceedings of 39th lunar and planetary science conference, Lunar and Planetary Science Institute, Texas, 10–14 Mar 2008

Grygorczuk J, Banaszkiewicz M, Kargl G, Kömle N, Ball A, Seweryn K (2009a) Use of hammering to determine cometary nucleus mechanical properties. In: Kargl G, Kömle NI, Ball AJ, Lorenz R (eds) Penetrometry in the solar system II. Austrian Academy of Sciences Press, Vienna

Grygorczuk J, Seweryn K, Wawrzaszek R, Banaszkiewicz M (2009b) Technological features in the new mole penetrator "KRET". In: Proceedings of ESAMATS: 13th European space mechanisms and tribology symposium, Vienna, 23–25 Sept 2009

Grygorczuk J, Seweryn K, Wawrzaszek R, Banaszkiewicz M, Rybus T (2009c) Moon's subsurface layers exploration by using "KRET" penetrator. AGH Drilling Oil Gas 26 (1–2):173–184

Gurgurewicz J, Rickman H, Królikowska M, Banaszkiewicz M, Grygorczuk J, Morawski M, Seweryn K, Wawrzaszek R (2010) Phobos investigations using the CHOMIK device (Phobos Sample Return mission). European Planetary Science Congress. http://adsabs.harvard.edu/abs/2010epsc.conf..683G. Accessed 9 Aug 2017

Haxel GB, Hedrick JB, Orris GJ (2002) Rare earth elements—critical resources for high technology. U.S. Geological Survey Report, Fact Sheet 087-02

Heiken GH, Vaniman DT, French BM (1991) Lunar sourcebook: a user's guide to the moon. Cambridge University Press, New York

Hoffman S, Andrews A, Watts A (2016) "Mining" water ice on mars: an assistant of ISRU options in support of future human missions. NASA. https://www.nasa.gov/sites/default/files/atoms/files/mars_ice_drilling_assessment_v6_for_public_release.pdf. Accessed 9 Aug 2017

Humphries M (2013) Rare earth elements: the global supply chain. Congressional Research Service. https://fas.org/sgp/crs/natsec/R41347.pdf. Accessed 2 Aug 2017

IISL (2015) Position paper on space resource mining. http://iislwebo.wwwnlss1.a2hosted.com/wp-content/uploads/2015/12/SpaceResourceMining.pdf. Accessed 9 Aug 2017

IISL (2016) Does international space law either permit or prohibit the taking of resources in outer space and on celestial bodies, and how is this relevant for national actors? What is the context, and what are the contours and limits of this permission or prohibition? International Institute of Space Law. http://iislweb.org/docs/IISL_Space_Mining_Study.pdf. Accessed 9 Aug 2017

ISECG (2013) Benefits steaming from space exploration. International Space Exploration Coordination Group. https://www.nasa.gov/sites/default/files/files/Benefits-Stemming-from-Space-Exploration-2013-TAGGED.pdf. Accessed 9 Aug 2017

ISECG Roadmap (2013) The global exploration roadmap. International Space Exploration Coordination Group. https://www.nasa.gov/sites/default/files/files/GER-2013_Small.pdf. Accessed 9 Aug 2017

ISECG (2014) The overview of the ISECG mission scenario. International Space Exploration Coordination Group

Jin YQ, Fa W (2007) Quantitative estimation of helium-3 spatial distribution on the lunar regolith layer. Icarus 190:15–23

JAXA (2015) JAXA's comprehensive scenario for space exploration (1st Iteration). International Space Exploration Promotion Team

Kömle NI, Kargl G, Seiferlin K, Marczewski W (2002) Measuring thermo-mechanical properties of cometary surfaces: in situ methods. Earth Moon Planet 90:269–282

Levine AL (1985) Commercialization of space: policy and administration issues. Public Adm Rev 45(5):562–569

Lewicki C, Diamandis P, Anderson E, Voorhees C, Mycroft F (2013) Planetary resources—the asteroid mining company. New Space 1(2):105–108

Liability Convention (1972) Convention on international liability for damage caused by space objects, UN GA Resolution 2777 (XXVI), annex, adopted on 29 Nov 1971, opened for signature on 29 Mar 1972, entered into force on 1 Sept 1972. United Nations Office for Outer Space Affairs

Luding S (2004) Molecular dynamics simulations of granular materials. In: Hinrichsen H, Wolf DE (eds) The physics of granular media. Wiley-VCH Verlag GmbH & Co. KGaA, Weinheim

Luxembourg to launch framework to support the future use of space resources (2017) Luxemburg Government. http://www.gouvernement.lu/5653386. Accessed: 4 Aug 2017

Meyers R (2015) The doctrine of appropriation and asteroid mining: incentivizing the private exploration and development of outer space. Or Rev Int Law 17(1):183–204

Moon 2020-2030: Outcome of the Symposium on a New Era of Human and Robotic Exploration (2016) Proceedings of Moon 2020-2030, ESA ESTEC, Noordwijk, 14–16 Dec 2015

Moon Agreement (1984) Agreement governing the activities of states on the moon and other celestial bodies, UN GA Resolution 34/68, annex, adopted on 5 Dec 1979, opened for signature on 18 Dec 1979, entered into force on 11 July 1984, Treaty Series 1363

NASA Fiscal Year (2017) Budget estimates. NASA. https://www.nasa.gov/sites/default/files/atoms/files/fy_2017_budget_estimates.pdf. Accessed 9 Aug 2017

NASA Technology Roadmaps (2015) TA 7: human exploration destination systems. NASA. https://www.nasa.gov/sites/default/files/atoms/files/2015_nasa_technology_roadmaps_ta_7_human_exploration_destination_final.pdf. Accessed 10 Aug 2017

Neal CR (2015) Moon 2020-2030 and the LEAG lunar exploration roadmap. Paper presented at the Moon 2020-2030, ESA ESTEC, Noordwijk, 14–16 Dec 2015

Outer Space Treaty (1967) Treaty on principles governing the activities of states in the exploration and use of outer space, including the moon and other celestial bodies, UN GA Resolution 2222 (XXI), annex, adopted on 19 December 1966, opened for signature on 27 Jan 1967, entered into force on 10 Oct 1967

PAP Nauka w Polsce (2017) Polskie firmy chcą rozwijać kosmiczne górnictwo. PAP. http://naukawpolsce.pap.pl/aktualnosci/news,413664,polskie-firmy-chca-rozwijac-kosmiczne-gornictwo.html. Accessed 26 July 2017

Pischel R (2017) The Russian Federal Space Programme. Papers presented at the 89th ESA international relations committee, ESA, Prague, 3–4 May 2017

Rare Earth Elements Profile (2010) Rare earth elements. British Geological Survey. http://nora.nerc.ac.uk/12583/1/Rare_Earth_Elements_profile.pdf. Accessed 10 Aug 2017

Rausch P (2016) The Luxembourg Government becomes a key shareholder of Planetary Resources, Inc., the U.S.-based asteroid mining company. Luxembourg Ministry of the Economy. http://www.businesswire.com/news/home/20161103005767/en/SpaceResources.lu-Luxembourg-Government-Key-Shareholder-Planetary-Resources. Accessed 27 July 2017

Registration Convention (1976) Convention on registration of objects launched into outer space, UN GA Resolution 3235 (XXIX), annex, adopted on 12 Nov 1974, opened for signature on 14 Jan 1975, entered into force on 15 Sept 1976. United Nations Office for Outer Space Affairs

Rescue Agreement (1969) Agreement on the rescue of astronauts, the return of astronauts and the return of objects launched into outer space, UN GA Resolution 2345 (XXII), annex, adopted on 19 Dec 1967, opened for signature on 22 Apr 1968, entered into force on 3 Dec 1968

Rickman H, Słaby E, Gurgurewicz J, Śmigielski M, Banaszkiewicz M, Grygorczuk J, Królikowska M, Morawski M, Seweryn K, Wawrzaszek R (2014) Chomik: a multi-method approach for studying phobos. Sol Syst Res 48(4):279–286

Ross SD (2001) Near-earth asteroid mining. space industry report. Control and Dynamical Systems Caltech, Pasadena. http://www.nss.org/settlement/asteroids/NearEarthAsteroidMining(Ross2001).pdf. Accessed 2 Aug 2017

Sacksteder K, Sanders GB (2007) In-situ resource utilization for lunar and mars exploration. In: Proceedings of 45th AIAA aerospace sciences meeting and exhibit, Reno, Nevada, 8–11 Jan 2007. https://doi.org/10.2514/6.2007-345

Sanders GB (2015) Space resource utilization: technologies and potential synergism with terrestrial mining. NASA. https://ntrs.nasa.gov/search.jsp?R=20150003499. Accessed 9 Aug 2017

Sanders GB, Linne DL, Star SO, Boucher D (2017) Leveraging terrestrial industry for utilization of space resources. In: Proceedings of planetary & terrestrial mining sciences symposium (PTMSS) and the space resources roundtable (SRR) 8th joint meeting, NASA and Canadian Institute of Mining, Montreal, 1–3 May 2017

Seweryn K (2016) The new concept of sampling device driven by rotary hammering actions. IEEE/ASME Trans Mechatron Syst 21(5)

Seweryn K, Wawrzaszek R, Grygorczuk J, Dąbrowski B, Banaszkiewicz M, Neal CR, Huang S, Kömle N (2008) Modelling of passive and active L-GIP thermal measurements in the lunar regolith. In: Proceedings of 39th lunar and planetary science conference, League City, Texas, 10–14 Mar 2008

Seweryn K, Banaszkiewicz M, Bednarz S, Gonet A, Grygorczuk J, Kuciński T, Rybus T, Rzyczniak M, Wawrzaszek R, Wiśniewski Ł, Wójcikowski M (2013) The experimental results of the functional tests of the mole penetrator KRET in different regolith analogues. In: Sasiadek J (ed) Aerospace robotics in GeoPlanet: Earth and Planetary Sciences. Springer, Berlin, pp 163–171

Seweryn K, Grygorczuk J, Wawrzaszek R, Banaszkiewicz M, Rybus T, Wisniewski Ł (2014a) Low velocity penetrators (LVP) driven by hammering action—definition of the principle of operation based on numerical models and experimental tests. Acta Astronaut 99:303–317

Seweryn K, Skocki K, Banaszkiewicz M, Grygorczuk J, Kolano M, Kucinski T, Mazurek J, Morawski M, Bialek A, Rickman H, Wawrzaszek R (2014b) Determining the geotechnical properties of planetary regolith using low velocity penetrometers. Planet Space Sci 99:70–83

Skonieczny K (2013) Lightweight robotic excavation. PhD thesis, Carnegie Mellon University

Skonieczny K, Delaney M, Wettergreen DS, Whittaker WL (2013) Productive lightweight robotic excavation for the Moon and Mars. J Aerosp Eng 27(4)

Skonieczny K, Moreland JL, Asnani VM, Creager CM, Inotsume H, Wettergreen DS (2014) Visualizing and analyzing machine-soil interactions using computer vision. J Field Robot 31 (5):820–836

Space Resource Exploration and Utilization Act of 2015 - H.R. 1508 (2015) Committee on science, space, and technology. https://www.congress.gov/bill/114th-congress/house-bill/1508/text. Accessed 9 Aug 2017

Spohn T, Seiferlin K, Hagermann A, Knollenberg J, Ball AJ, Banaszkiewicz M, Benkhoff J, Gadomski S, Gregorczyk W, Grygorczuk J, Hlond M, Kargl G, Kührt E, Kömle N, Krasowski J, Marczewski W, Zarneck JC (2007) Mupus—a thermal and mechanical properties probe for the Rosetta Lander Philae. Space Sci Rev 128(1–4):339–362

Tanaka S (2017) Japan plans to land astronauts on Moon around 2030. http://www.asahi.com/ajw/articles/AJ201706290030.html. Accessed 4 Aug 2017

Tronchetti F (2009) The exploitation of natural resources of the Moon and other celestial bodies. Brill, Nijhoff

Tsuda Y, Yoshikawa M, Abe M, Minamino H, Nakazawa S (2013) System design of the HAYABUSA 2—asteroid sample return mission to 1999 JU3. Acta Astronaut 91:356–362

Ulivi P, Harland DM (2015) Robotic exploration of the solar system Part IV: the modern era 2004–2013. Springer, Berlin

United Nations Convention on the Law of the Sea Division for Ocean Affairs and the Law of the Sea (1982) Office of legal affairs, United Nations

Velocci A (2012) Commercialization in space. Harvard International Review. http://hir.harvard.edu/article/?a=2921. Accessed 4 Aug 2017

Williams DR (2005) Soviet lunar missions. NASA Goddard Space Flight Center. https://nssdc.gsfc.nasa.gov/planetary/lunar/lunarussr.html. Accessed 8 Aug 2017

Wilson A (2005) ESA Achievements: more than thirty years of pioneering space activity. ESA ESTEC, Noordwijk

Zacny K, Bar-Cohen Y, Brennan M, Briggs G, Cooper G, Davis K, Dolgin B, Glaser D, Glass B, Gorevan S, Guerrero J, McKay C, Paulsen G, Stanley S, Stoker C (2008) Drilling systems for extraterrestrial subsurface exploration. Astrobiology 8(3):665–706

Zelenyi LM (2016) Along the thorny pass of Space Science. 50-years-journey. In: Space Research Institute in Times of Change. Glimpses of the past and visions of the future. Proceedings of the international forum "Space Science: Yesterday, Today and Tomorrow", Space Research Institute of the Russian Academy of Sciences (IKI RAN), Moscow, 30 Sept–2 Oct 2015

Space Mechatronics and Space Robotics Patent Inventions; the Way to Protect the Space Heritage in the Space Research Centre, Institute of the Polish Academy of Sciences

Marta E. Wachowicz and Marek Bury

1 Introduction

Space Research Centre of the Polish Academy of Sciences continuously develops and strictly implements Intellectual Property Rights Management Policy (IPRMP). Its aim is to identify, report, secure, and deploy patents, utility models, copyrights, designs and trademarks consistent with the Institute mission of research and technology development. IPRMP is a part of intellectual property strategy that will contribute to both protection of the space heritage and the newest ideas, and to the efficient commercialization process in the Institute on the European space market. It was noted that intensified patent protection results in surprisingly high motivation of the researchers towards starting up technology based on private entities to obtain licenses and offer new products to the industry. For this reason intellectual property protection contributes to the generation of space industry sector in Poland, which earlier barely existed. Implicitly, it will increase significantly the patent protection of Polish companies in the space sector and thus enables equal opportunities in competition with foreign space entities. Space Research Centre of the Polish Academy of Sciences is responsible for managing the intellectual property assets of the institute for the public good and takes the financial risk in patenting and licensing intellectual property.

M. E. Wachowicz (✉)
Polish Space Agency, Powsińska 69/71, 02-903 Warsaw, Poland
e-mail: marta.wachowicz@polsa.gov.pl

M. Bury
Bury & Bury Kancelaria Patentowa, Słowackiego 5/13 lok. 111, 01-592 Warsaw
Poland
e-mail: marek.bury@bnb-ip.com.pl

J. Sasiadek, *Aerospace Robotics III*, GeoPlanet: Earth and Planetary Sciences,
https://doi.org/10.1007/978-3-319-94517-0_11

2 SRC PAS Space Heritage

SRC PAS is an interdisciplinary research institute of the Polish Academy of Sciences established to develop space activities in Poland. It is the only institute in Poland whose activity is fully dedicated to the research into extraterrestrial space, the Solar System and the Earth using space technology and satellite techniques. The main activities are connected with research related to the satellite experiments on space physics and physical processes on planets and on Earth, participating in the international space missions and research programs, and constructing space instruments and satellite components for research purposes, as well as developing space technologies, especially in robotics and mechatronics field. Last but not least research activity of SRC PAS is application of satellite techniques for terrestrial telecommunication, navigation and Earth observation.

Participation in space missions has given SRC PAS's engineers and specialists active knowledge of the difficult art of building complex structures, carried out in accordance with the very demanding procedures of European Space Agency (ESA). Experience of participation in huge number of missions accumulates into know-how and human resources of engineers who learned their skills solving real problems and constructing working space devices.

Instruments designed and built in SRC PAS are unique on a global scale and consist of a significant contribution to world space heritage in the field of mechatronics. In 40 years of its history, SRC PAS has acquired a considerable experience in building instruments, and in some areas reached a very high level of qualification of space technology such as TRL6, TRL7, TRL8. Participation in the international missions and participation in the most prestigious scientific experiments stress out the importance of the Institute's experience. SRC PAS scientists gained an experience of participation in the international research groups and consortia, and especially in the four main areas of science and engineering:

- Construction of optical instruments including optical instruments for missions: Mars Express, Venus Express, Chandrayaan, Bepi-Colombo,
- Designing of electronic instruments for FHLCU in Herschel,
- Designing mechatronic devices and sensors: CASSINI/Huygens (experiment SSP) Rosetta mission to comet Churyumov-Gerasimenko (experiment MUPUS) and Obstanovka experiment on the International Space Station, mission Phobos Grunt (CHOMIK),
- Preparing software to support instruments and experiments for mission INTEGRAL, IBIS experiments and JEM-X, mission HERSCHEL, an experiment HiFi, DEMETER and TARANIS, IBEX, Obstanovka and ASIM.

It is a truth quite universally acknowledged that space dedicated mechanisms have to be lightweight and durable, particularly extremely reliable. It is a simple consequence of high cost of placing devices in outer space and barely any possibility of servicing them. Simultaneously, space equipment is subjected to unique and extreme conditions. To main factors that might affect space mechanisms belong high vacuum, extremely low temperatures (even −160 °C) and micro-gravity conditions. The development of reliable mechanisms working in space requires very good knowledge of that environment, assessment of risks, and the use of modern technologies and extensive testing. Lack of atmosphere results in non-existent convection of space. Cooling of space mechanisms is substantially based on radiation of heat. Therefore despite ambient temperature so extremely low that it changes mechanical parameters of metals overheating still may be a problem. This effect combined with possibility of exposition to the radiation of Sun extends effective range of operating temperatures for the space equipment to more than 200°.

A particular speciality of SRC PAS is construction of ground penetrators. Penetrators are objects to penetrate the small bodies of the Solar System such as asteroids and comets, as well as large bodies penetrating planets and moons, have been used for years for scientific and technical purposes. For penetration of any celestial bodies are essential mechanical properties of the soil: its structure, strength, compactness and porosity. Space penetrators are almost indispensable in following research activities: chemical and mineralogical analyses of in situ thermal (temperature profile, thermal conductivity), mechanical ground, electrical and magnetic or penetration radar (GPR). Penetrators are also used for sampling of soil from different depths and delivering them to the surface.

The high level of scientific and innovative solutions is presented by the three selected penetrators—MUPUS, KRET and CHOMIK (Rickman et al. 2014). SRC PAS participated in the international experiment, which was performed for the purpose of MUPUS (Multi Purpose Sensor for Surface and Subsurface science) (Grygorczuk et al. 2007). Polish contribution in this experiment included the scientific and technical aspects of the project, (including) such as manufacturing and technical qualification of the instrument. This instrument is actually a multifunctional robot. The requirements of MUPUS construction were extremely difficult, since devices are meant to work under extreme conditions, i.e. in a vacuum, with great range and temperature gradients, with resistance to huge overload (even required 1000 g), and at the same time with the expectation of a good reliability after several years of space travel (Spohn et al. 2007).

Light-weight manipulator based on tubular booms technology is a further significant achievement of SRC PAS. Tubular boom was used for the deployment of MUPUS penetrator on board of the ESA's Rosetta lander Philae, the Ultra-Light Planetary Manipulator, that is also based on the same booms, was constructed. (Annual Report SRC PAS 2012).

It should be underlined that the requirements in space sector are very high; under specific conditions in space the products must be characterized by unconventional engineering solutions, especially in the field of mechatronics, optics, electronics, engineering, physics.

On the one hand, variety of technologies and space market, on the other unique knowledge proved in the space environment, require special treatment in the field of intellectual protection and patent policy.

3 Mechatronics and Robotic Inventions; Patent Protection

Portfolio of the Institute includes a huge number of unique designs engineering masterpieces and instruments. However, the number of patents and patents application directed to protecting this portfolio had been surprisingly low. Due to the high complexity, the requirements of the space industry and limited number of entities recognised as reliable space partners it is difficult to imitate the space oriented achievements. SRC PAS portfolio had been well known and recognised and there was no need to protect the intellectual property assets using patents. At present the situation is changed. New challenges (ESA membership, development of space sector in Poland and academic entrepreneurship) and sudden increase of interest of industry in space applications forced the institute to revise the policies and decisions regarding patent protection.

Several factors influence patent protection decision. First of all, patent for space equipment gives monopoly for the entity to be commercial provider of such equipment for space missions. Patent application can also be treated as background intellectual property right, what is specially important for ESA projects and tenders. Secondly, patent is still recognised as being prestigious. On one hand, it shows the financial engagement of the owner, on the other not only are patent applications and patents expensive, but they teach competitors a lot about the strategies, technologies, and plans for the future. Thirdly, it is important to understand all the features of a patent law, especially concerning the ownership regulation. Patent law has provisions for rewarding the inventors and enforces establishment of codified relations applicant-inventor. It is worth mentioning that patent law is well suited for technology transfer. Hence it facilitates reuse of space inventions in terrestrial applications. Therefore, applying for patent protection supplies space sector in Poland with findings due to licencing of the intellectual property rights to Polish Small and Medium-sized Enterprises (SME), whether they are space oriented or not.

SME may be interested in starting a new space business, expanding an existing business (extending territory or changing the focus) and thereby improving its market position. In many situations, licensing of intellectual property rights is an effective tool for achieving business goals, or even it is the only possibility to compete for international contracts or apply for ESA tenders. In the international context, a formal licensing agreement of space technology is reasonable only if the intellectual property right to be licensed has a territorial scope that covers all countries involved (Should there be otherwise licensor would not only be able to enforce exclusive right against third parties and licensee might not be willing to accept situation in which he would be obliged to pay for the technology that is free to other entities). In the view of the above, Institute has to carefully select geographical scope of protection and design it to meet the needs of the business and future development plan.

The Institute is a beneficiary of the Operational Programme Innovative Economy, the main aim of the project *Protection of space technology through patent applications key inventions in the field of mechatronics and robotics space* is to increase the number of robotics and mechatronics patent applications and due to licencing of the intellectual property rights develop polish space sector. The aim of the project is also to use the intellectual potential of SRC PAS to implement developed technologies for space industry and terrestrial applications. As a result of the project in the years 2013–2015 SRC PAS has filed 8 Polish patent applications, 4 international (Patent Cooperation Treaty-PCT) applications, 3 European patent applications (European Patent Convention-EPC) and 2 USPTO (The United States Patent and Trademark Office) applications. All inventions were related to space robotics and mechatronics, for example PCT/IB2014/062651: "Mechanism, method and escapement wheel for controlling rotational motion" or PCT/IB2014/064589 "Clamp mechanism, locking system and method of operating a reconfigurable manipulator".

The result shows the importance of creating and managing a SRC PAS portfolio of own protected solutions, including the choice of an appropriate form and territorial scope of protection for the results of R&D works in the field of space robotics and mechatronics.

4 Space Technology Transfer

Space technology transfer is different than in other branches of R&D, due to the very high investment costs, special requirements for space technology, the need for dedicated infrastructure and highly specialized engineers. The most efficient innovations often derive from using existing technologies for purposes unrelated to their original applications.

A very important and promising channel of commercialization of research results and achieving the benefits of intellectual property protection is the use of space research knowledge to terrestrial, non-space applications. Products designed for space exploration fulfil the construction requirements that are adequate to the extreme conditions in space, such as very low temperatures and significant temperature gradient depending on the Sun position, vacuum, different than from the Earth's gravitational field, cosmic dust pollution, strong cosmic radiation. Any instruments, devices and elements are designed, tested and assembled with extreme precision, using specialized and dedicated hardware, and under very strictly controlled conditions. Instruments exploring interplanetary space are exposed to many dangers, such as collision with minor bodies, strong magnetic storms, increased solar activity, solar wind streams. Therefore, space engineering requires the development of very high-performance products. Space technology, components or applications constitute a huge potential for terrestrial applications.

Space technologies are the reservoir of innovation; numerous interesting applications of technology or materials that had initially been developed for space have proved to be successful in common life application. This phenomenon contributes to the need of obtaining patent protection for space invention. If they are protected, they are easier to commercialized by true entitled entity (Institute) and makes it easier to establish so-called spin-offs. Patent protection increases the chances of commercialization on the European market, also creates negotiating position, clearly participates in new missions and building space experiments for the European Space Agency. Patent protection of robotic and mechatronics inventions can result in business opportunity, taking into account two major groups of recipients: the European space market with the European Space Agency and high technology companies sector, involving the production or use of advanced or sophisticated devices and new materials.

The space European market is defined by the prime or subcontractor. The accession to the ESA has profound economic justification, and Polish industry can rely on contracts with a value of at least 80% of Polish membership fee. An obstacle to the use of this opportunity, however, may face a dramatic superiority of foreign entities in terms of industrial property protection. There is a lack of Polish companies among the leading applicants for patenting inventions.

Implementation of coherent patent policy in the field of space robotics can start the cycle change and allow the use of chance to work with the ESA. According to the document, the European Space Agency setting out the vision and challenges for the XXI century (Agenda 2015-ESA BR-303), ESA recognizes robotics as one of the key technologies to maintain its leading role in space exploration. Potentially, there is a possibility of the market share in the following ESA optional programs associated with mechatronics and robotics space, or fields related to the project ExoMars, Lunar Lander, or Space Situational Awareness, including—Space Surveillance and Tracking (tracking technologies in orbit, deorbiting technologies, space debris, orbital robotics). According to the priorities of ESA, development of robotic technology is one of the most important competencies which agency should develop in the next years. Polish entities will be able to develop technologies that will in the future be used in all missions and ESA programs ranging from space-flight (preparation of robots to cooperate with humans) to orbital robotics and building the landers. As part of this program, the following technologies could be developed: advanced power systems, power management systems, navigation, mechanics and electronics, advanced mechanical construction, suspension and drive systems, optical components and autonomy software.

5 Conclusions

There is a need for constant improvement of the Intellectual Property Rights Management Policy in SRC PAS. Intellectual property rights are a compromise, representing the search for balance between making all knowledge available within the public domain and granting ownership of valuable discoveries to the Institute. Reaching an appropriate balance requires therefore continuous intellectual property management. Issues related to space robotics and mechatronics should be addressed with particular care. Issues related to space robotics and mechatronics should be addressed with particular care. The process of patent portfolio strengthening should focused on inventing around its core capabilities and for complementary intellectual property. Taking into account the plans of SRC PAS for subsurface explorations and participation in future lunar or Mars space mission, an enhancement of core IP-driven products, strictly connected with space heritage of SRC PAS, is crucial. Strategic patenting to protect a core idea starts with building a patent portfolio around the core robotic or mechanic technology that covers not only what institute does, but also the final product or instrument, service or market alternatives that could allow to present the SRC PAS achievements at European space market. Effective patent protection can also stimulate research in these fields, like space robotics or mechatronics and often appear to be a key requirement for raising capital for further development and experiments. A patent portfolio enables the institute to potentially recoup development costs and obtain a return of investment in the development of the patented technology. Patent strategy following the space development trends can also generate revenue from the licensing patents. Additionally, institute valuation relies greatly on an intellectual assets, such as, patents.

Acknowledgements IP strategy policy and patent applications reported in this paper were supported by Project *Protection of space technology through key patent inventions in the field of mechatronics and robotics space* implemented under the Operational Programme Innovative Economy 2007–2013, Grant No. UDA-POIG.01.03.02-00-011/12-00.

References

Agenda 2015-ESA BR-303 (29.11.2011), ESA

Annual Report SRC PAS (2012) SRC PAS, Warszawa

Grygorczuk J, Banaszkiewicz M, Seweryn K, Spohn T (2007) MUPUS insertion device for Rosetta mission. J Telecommun Inf Technol 1:50–53

Rickman H, Słaby E, Gurgurewicz J, Śmigielski M, Banaszkiewicz M, Grygorczuk J, Morawski M, Seweryn K, Wawrzaszek R (2014) CHOMIK: a multi-method approach for studying Phobos solar system research. Sol Syst Res 48:279–286

Spohn T, Seiferlin K, Hagermann A, Knollenberg J, Ball AJ, Banaszkiewicz M, Benkhoff J, Gadomski S, Grygorczuk J, Hlond M, Kargl G, Kührt E, Kömle N, Marczewski W, Zarnecki JC (2007) MUPUS—a thermal and mechanical properties probe for the Rosetta lander PHILAE. Space Sci Rev 128:339–362

9783319945163